GOOD NATURED

GOOD NATURED

The Origins of
Right and Wrong in
Humans and Other Animals

Frans de Waal

Harvard University Press

Cambridge, Massachusetts • London, England

Library of Congress Cataloging-in-Publication Data

Waal, F. B. M. de (Frans B. M.), 1948–
Good Natured : the origins of right and wrong in humans
and other animals / Frans de Waal.
p. cm.
Includes bibliographical references and index.
ISBN 0-674-35660-8 (alk. paper) (cloth)
ISBN 0-674-35661-6 (pbk.)
1. Ethics, Evolutionary. 2. Animal behavior.
3. Human behavior.
4. Ethics. I. Title.
BJ1335.W33 1996
599'.052'-4—dc20 95-46032

à ma Cattie

CONTENTS

PROLOGUE

In addition to being human, we pride ourselves on being *humane*. What a brilliant way of establishing morality as the hallmark of human nature—by adopting our species name for charitable tendencies! Animals obviously cannot be human; could they ever be humane?

If this seems an almost-rhetorical question, consider the dilemma for biologists—or anyone else adopting an evolutionary perspective. They would argue that there must at some level be continuity between the behavior of humans and that of other primates. No domain, not even our celebrated morality, can be excluded from this assumption.

Not that biologists have an easy time explaining morality. Actually, there are so many problems with it that many would not go near the subject, and I may be considered foolish for stepping into this morass. For one thing, inasmuch as moral rule represents the power of the community over the individual, it poses a profound challenge to evolutionary theory. Darwinism tells us that traits evolve because their bearers are better off with them than without them. Why then, are collective interests and self-sacrifice valued so highly in our moral systems?

Debate of this issue dates back a hundred years, to 1893 when

Thomas Henry Huxley gave a lecture on "Evolution and Ethics" to a packed auditorium in Oxford, England. Viewing nature as nasty and indifferent, he depicted morality as the sword forged by *Homo sapiens* to slay the dragon of its animal past. Even if the laws of the physical world—the cosmic process—are unalterable, their impact on human existence can be softened and modified. "The ethical progress of society depends, not on imitating the cosmic process, still less in running away from it, but in combating it."[1]

By viewing morality as the antithesis of human nature, Huxley deftly pushed the question of its origin outside the biological realm. After all, if moral conduct is a human invention—a veneer beneath which we have remained as amoral or immoral as any other form of life—there is little need for an evolutionary account. That this position is still very much with us is illustrated by the startling statement of George Williams, a contemporary evolutionary biologist: "I account for morality as an accidental capability produced, in its boundless stupidity, by a biological process that is normally opposed to the expression of such a capability."[2]

In this view, human kindness is not really part of the larger scheme of nature: it is either a cultural counterforce or a dumb mistake of Mother Nature. Needless to say, this view is extraordinarily pessimistic, enough to give goose bumps to anyone with faith in the depth of our moral sense. It also leaves unexplained where the human species can possibly find the strength and ingenuity to battle an enemy as formidable as its own nature.

Several years after Huxley's lecture, the American philosopher John Dewey wrote a little-known critical rejoinder. Huxley had compared the relation between ethics and human nature to that between gardener and garden, where the gardener struggles continuously to keep things in order. Dewey turned the metaphor around, saying that gardeners work as much *with* nature as against it. Whereas Huxley's gardener seeks to be in control and root out whatever he dislikes, Dewey's is what we would today call an organic grower. The successful gardener, Dewey pointed out, creates conditions and introduces plant species that may not be normal for this particular plot of land "but fall within the wont and use of nature as a whole."[3]

I come down firmly on Dewey's side. Given the universality of moral systems, the tendency to develop and enforce them must be an integral part of human nature. A society lacking notions of right and wrong is about the worst thing we can imagine—if we can imagine it at all. Since we are moral beings to the core, any theory of human behavior that does not take morality 100 percent seriously is bound to fall by the wayside. Unwilling to accept this fate for evolutionary theory, I have set myself the task of seeing if some of the building blocks of morality are recognizable in other animals.

Although I share the curiosity of evolutionary biologists about *how* morality might have evolved, the chief question that will occupy us here is *whence* it came. Thus, after due attention in this book's first chapter to theories of evolutionary ethics, I will move on to more practical matters. Do animals show behavior that parallels the benevolence as well as the rules and regulations of human moral conduct? If so, what motivates them to act this way? And do they realize how their behavior affects others? With questions such as these, the book carries the stamp of the growing field of *cognitive ethology:* it looks at animals as knowing, wanting, and calculating beings.

As an ethologist specialized in primatology, I naturally turn most often to the order of animals to which we ourselves belong. Yet behavior relevant to my thesis is not limited to the primates; I include other animals whenever my knowledge permits. All the same, I cannot deny that primates are of special interest. Our ancestors more than likely possessed many of the behavioral tendencies currently found in macaques, baboons, gorillas, chimpanzees, and so on. While human ethics are designed to counteract some of these tendencies, in doing so they probably employ some of the others—thus fighting nature with nature, as Dewey proposed.

Because my goal is to make recent developments in the study of animal behavior accessible to a general audience, I draw heavily on personal experience. Interacting with animals on a daily basis, knowing each of them individually, I tend to think in terms of what I have seen happen among them. I am fond of anecdotes, particularly those that capture in a nutshell social dynamics that would take a thousand words to explain. For the same reason, this book is liberally illustrated with photographs (which, unless otherwise specified, are mine).

At the same time, vignettes do not constitute scientific proof. They tease the imagination and sometimes hint at striking capacities, yet cannot demonstrate them. Only repeated observations and solid data allow us to compare alternative hypotheses and arrive at firm conclusions. The study of animal behavior is conducted as much behind the computer as at the observation site. Over the years, my students and I have recorded large amounts of systematic data on group-living primates, mostly in outdoor enclosures at zoos and research institutions. In addition, a host of colleagues have been assiduously working

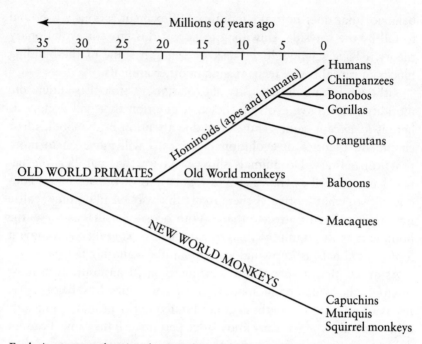

Evolutionary tree showing the main branches of the primate order: the New World monkeys, the Old World monkeys, and the hominoid lineage that produced our own species. This diagram reflects recent advances in DNA analysis that place the African apes (gorillas, chimpanzees, and bonobos) much closer to humans than previously thought.

on related issues, both in the laboratory and in the field. In an attempt to integrate these approaches, at least half of the material presented herein concerns research by others.

Because my writing alternates between stories, theories, and hard-won data, it risks blurring the line between fact and speculation. To help readers distinguish between the two and explore certain topics at greater length, the book includes technical notes as well as an extensive bibliography. Although by no means exhaustive, this additional material makes clear that rigorous scientific methods can be and are being applied to some of the questions at hand.

Western science seems to be moving away from a tidy, mechanistic worldview. Aware that the universe is not necessarily organized along logically consistent lines, scientists are—ever so reluctantly—beginning to allow contradictions. Physicists are getting used to the idea that energy may be looked at as waves but also as particles, and economists that free-market economies can be beaten at their own game by guided economies such as that of the Japanese.

In biology, the very same principle of natural selection that mercilessly plays off life forms and individuals against one another has led to symbiosis and mutualism among different organisms, to sensitivity of one individual to the needs of another, and to joint action toward a common goal. We are facing the profound paradox that genetic self-advancement at the expense of others—which is the basic thrust of evolution—has given rise to remarkable capacities for caring and sympathy.

This book tries to keep such conflicting thoughts simultaneously aloft. The one is not easily reduced to the other, although attempts have been made, most prominently the proposition that deep down, concern for others always remains selfish. By denying the existence of genuine kindness, however, these theories miss out on the greater truth emerging from a juxtaposition of genetic self-interest and the intense sociality and conviviality of many animals, including ourselves.

Instead of human nature's being either fundamentally brutish or fundamentally noble, it is both—a more complex picture perhaps, but an infinitely more inspiring one.

1

DARWINIAN DILEMMAS

Be warned that if you wish, as I do, to build a society in which individuals cooperate generously and unselfishly towards a common good, you can expect little help from biological nature. Let us try to *teach* generosity and altruism, because we are born selfish.

Richard Dawkins[1]

Why should our nastiness be the baggage of an apish past and our kindness uniquely human? Why should we not seek continuity with other animals for our 'noble' traits as well?

Stephen Jay Gould[2]

Famous in her country as the star of several nature documentaries, Mozu looks like any other Japanese monkey except for missing hands and feet and an arresting countenance that appears to reflect lifelong suffering. She roams the Shiga Heights of the Japanese Alps on stumpy limbs, desperately trying to keep up with more than two hundred healthy group mates. Her congenital malformations have been attributed to pesticides.

When I first visited Jigokudani Park in 1990, Mozu was already eighteen years old—past prime for a female macaque. She had successfully raised five offspring, none of whom showed abnormalities. Given the extended period of nursing and dependency of primate young, no one would have dared to predict such a feat for a female who must crawl over the ground, even in midwinter, to stay with the rest. While the others jump from tree to tree to avoid the ice and snow covering the forest floor, Mozu slips and slides through shoulder-high snow with an infant on her back.

One thing that the monkeys in Jigokudani Park have in their favor is hot-water springs, in which they temporarily escape from the glacial temperatures, grooming one another amid clouds of steam. Another factor that makes life easier is food provisioning. Modest amounts of soybeans and apples are distributed twice daily at the park. Caretakers say they give Mozu extra food and protect her when she encounters competition from other monkeys. They try to make up for the trouble she has obtaining food, yet stress that Mozu does not dally at the feeding site. She is really part of the troop. Like the rest, she spends most of her time in the mountain forest, away from people.

Survival of the Unfittest

My first reaction to Mozu was one of awe: "What a will to live!" The connection with morality came later, when I heard how much paleontologists were making of the occasional survival into adulthood of Neanderthals and early humans afflicted with dwarfism, paralysis of the limbs, or inability to chew. With exotic names such as Shanidar I, Romito 2, the Windover Boy, and the Old Man of La Chapelle-Aux-Saints, the fossil remains of a handful of cripples were taken to mean that our ancestors supported individuals who could contribute little to the community. Survival of the weak, the handicapped, the mentally retarded, and others who must have posed a burden was depicted as the first appearance on the evolutionary scene of compassion and moral decency. Cavemen turned out to be communitarians under the skin.

Accepting this logic, should we not also include Mozu's survival as an example of moral decency? One might counter that the artificial food provisioning at Jigokudani Park disqualifies her, since we do not know if she would have made it without the extra food. Moreover, if active community support is our criterion, Mozu can be eliminated right away because there is no shred of evidence that other monkeys have ever gone out of their way to assist her in her monumental struggle for existence.

Exactly the same arguments have been raised against the Shanidars and Romitos of the human fossil record. According to K. A. Dettwyler, an anthropologist, it is possible that these individuals lived in rich environments in which the sharing of resources with a few impaired community members posed no problem. In return, the handicapped individuals may have made themselves useful by collecting firewood, baby-sitting, or cooking. Dettwyler also argues that there is a wide gap between mere survival and being treated well. She describes cultures in which mentally retarded people are stoned, beaten, and jeered at for public amusement, or in which people afflicted with polio do not receive any special consideration ("adult women crawled on hands and knees with children tied to their backs").[3] As for Western society, we need only think of the filthy asylums of the not-too-distant past, and the chained existence of the insane, to realize that survival does not necessarily imply humane conditions.

Without knowing the precise similarities and differences between Mozu and the human fossils, I do not think these fossils prove moral decency any more than does Mozu's survival. Only a relatively tolerant attitude toward the handicapped can be inferred in both cases. Mozu is certainly well accepted by her group mates, a fact that may have contributed to her survival. If what happened in 1991 is any measure, Mozu may even enjoy a special level of tolerance.

In the spring of that year, the troop of monkeys at Jigokudani had grown so large that it split in half. As usual during fissioning, the dividing line followed the backbone of macaque society, the matrilineal hierarchy (female kin are closely bonded and united in their battles with nonkin, the result being a social order based on matrilineal descent). One piece of the troop consisted of a few dominant matriarchs and their families; the other included subordinate matriarchs and their families. Being of low rank, Mozu and her offspring ended up in the second division.

According to Ichirou Tanaka, a Japanese primatologist who has worked at the park for years, the fission posed a serious problem for Mozu. The dominant division began to claim the park's feeding site for itself, aggressively excluding all other monkeys. Faced with this situation, Mozu made a unique decision. Whereas female macaques normally maintain lifelong bonds of kinship, Mozu ignored the ties with her offspring and began making overtures to individuals in the dominant division. Despite occasional attacks on her, she stayed at the periphery, seeking contact with age-peers, females with whom she had grown up nineteen years before. She made repeated attempts to groom them (without fingers, Mozu's rather clumsy grooming still served to initiate contact). Eventually her peers began to accept her presence, and to return Mozu's grooming. Mozu is now well integrated into the dominant troop, once again enjoying the feeding site, yet having paid for this advantage with permanent separation from her kin.

In no society worthy of the name do the members lack a sense of belonging and a need for acceptance. The ability and the tendency to construct such associations, and to seek security within them, are products of natural selection found in members of species with better survival chances in a group than in solitude. The advantages of group life can be manifold, the most important being increased chances to find food, defense against predators, and strength in numbers against competitors. For example, it may be of critical importance during a drought to have older individuals around who can lead the group to an almost-forgotten waterhole. Or, during periods of heavy predation all eyes and ears count, especially in combination with an effective warning system. Each member contributes to and benefits from the group, although not necessarily equally or at the same time.

Mozu's case teaches us that even though primate groups are based on such give-and-take contracts, there is room for individuals with little value when it comes to cooperation. The cost to the others may be negligible, but their inclusion is remarkable, given the realistic alternative of ostracism.

Noting that Japanese monkeys can be quite aggressive, at times demonstrating what he calls murderous intent, Jeffrey Kurland described the following concerted action against a particular matriline at a site far from Jigokudani.

A female of the top matriline started a fight with a low-ranking female named Faza-71. The attacker and her supporters (a sister, a brother, and a niece) made so much noise that the alpha male (the troop's most dominant male) was attracted to the scene. By the time he arrived, Faza-71 was high in a tree, a position from which she was forced to jump 10 meters to the ground when the male climbed up and cuffed her. Fleeing from her pursuers, Faza-71 saw no escape other than an icy, fast-streaming river. Her attackers wisely stayed on land, but for a long time prevented the frantically swimming Faza-71 from coming back on the riverbank. In the meantime Faza-71's family, powerless to help, fled over a dam across the river.

But for a small pile of sand under a chilly waterfall, Faza-71 would have drowned. Bleeding and apparently in shock, she waited to join her family until the attackers had dispersed. The entire encounter lasted less than half an hour; but it took more than a week for Faza's matriline to rejoin the troop, and many months for them to relax in the presence of the dominant matriline.[4]

Biologicizing Morality

Social inclusion is absolutely central to human morality, commonly cast in terms of how we should or should not behave in order to be valued as members of society. Immoral conduct makes us outcasts, either here and now or—in the beliefs of some people—when we are turned away from the gates of heaven. Universally, human communities are moral communities; a morally neutral existence is as impossible for us as a completely solitary existence. As summed up by Mary Midgley, a philosopher, "Getting right outside morality would be rather like getting outside the atmosphere."[5] Human morality may indeed be an extension of general primate patterns of social integration, and of the adjustment required of each member in order to fit in. If so, the broadest definition of this book's theme would be as an investigation into how the social environment shapes and constrains individual behavior.

No doubt some philosophers regard morality as entirely theirs. The claim may be justifiable with regard to the "high end" of morality: abstract moral rules can be studied and debated like mathematics, almost divorced from their application in the real world. According to child psychologists, however, moral reasoning is constructed upon much simpler foundations, such as fear of punishment and a desire to conform. In general, human moral development moves from the social to the personal, from a concern about one's standing in the group to an autonomous conscience. While the early stages hardly seem out of reach of nonhuman animals, it is impossible to determine how close they get to the more rational, Kantian levels. Reliable nonverbal signs of thought in humans do not exist, and the indicators that we sometimes do use (staring into the distance, scratching the head, resting the chin on a fist) are commonly observed in anthropoids. Would an extraterrestrial observer ever be able to discern that humans ponder moral dilemmas, and if so, what would keep that observer from arriving at the same conclusion for apes?

Biologists take the back door to the same building that social scientists and philosophers, with their fondness for high-flung notions, enter through the front door. When the Harvard sociobiologist E. O. Wilson twenty years ago proclaimed that "the time has come for ethics to be removed temporarily from the hands of philosophers and biologicized,"[6] he formulated the same idea a bit more provocatively. My own feeling is that instead of complete reliance on biology, the best way to generate fresh air is simultaneously to open both front and back doors. Biologists look at things in a rather functional light; we always wonder about the utility of a trait, on the assumption that it would not be there if it did not serve some purpose. Successful traits contribute to "fitness," a term that expresses how well adapted (fitted) an individual is to its environment. Still, emphasis on fitness has its limitations. These are easily recognized when paleontologists hold up the fossil remains of an ancestor who could barely walk, declaring it a defining moment in human prehistory when the unfit began to survive.

To understand the depth of these limitations, one need only realize the influence of Thomas Malthus' essay on population growth that appeared at the beginning of the nineteenth century. His thesis was that populations tend to outgrow their food supply and are cut back automatically by increased mortality. The idea of competition within the *same* species over the *same* resources had immediate appeal to Charles Darwin, who read Malthus; it helped bring his Struggle for Existence principle into focus.

Sadly, with these valuable insights came the burden of Malthus' political views. Any help one gives the poor permits them to survive and propagate, hence negates the natural process according to which these unfortunates are supposed to die off. Malthus went so far as to claim that if there is one right that man clearly does *not* possess, it is the right to subsistence that he himself is unable to purchase with his labor.[7]

Although Darwin appears to have struggled more with the moral implications of these ideas than most of his contemporaries, he could not prevent his theory from being incorporated into a closed system of thought in which there was little room for compassion. It was taken to its extreme by Herbert Spencer in a grand synthesis of sociology, political economy, and biology, according to which the pursuit of self-interest, the lifeblood of society, creates progress for the strong at the expense of the inferior. This convenient justification of disproportionate wealth in the hands of a happy few was successfully exported to the New World, where it led John D. Rockefeller to portray the expansion of a large business as "merely the working-out of a law of nature and a law of God."[8]

Given the popular use and abuse of evolutionary theory (comparing Wall Street to a Darwinian jungle, for example), it is not surprising that in the minds of many people natural selection has become synonymous with open, unrestricted competition. How could such a harsh principle ever explain the concern for others and the benevolence encountered in our species? That a reason for such behavior does not follow readily from Darwin's theory should not be held against it. In the same way that birds and airplanes appear to defy the law of gravity yet are fully subjected to it, moral decency may appear to fly in the face of natural selection yet still be one of its many products.

Altruism is not limited to our species. Indeed, its presence in other species, and the theoretical challenge this represents, is what gave rise to *sociobiology*—the contemporary study of animal (including human) behavior from an evolutionary perspective. Aiding others at a cost or risk to oneself is widespread in the animal world. The warning calls of birds allow other birds to escape a predator's talons, but attract attention to the caller. Sterile castes in social insects do little else than serve food to the larvae of their queen or sacrifice themselves in defense of their colony. Assistance by relatives enables a breeding pair of jays to fill more hungry mouths and thus raise more offspring than otherwise possible. Dolphins support injured companions close to the surface in order to keep them from drowning. And so on.

Should not a tendency to endanger one's life for someone else be quickly weeded out by natural selection? It was only in the 1960s and 1970s that satisfactory explanations were proposed. According to one theory, known as *kin selection*, a helping tendency may spread if the help results in increased survival and reproduction of kin. From a genetic perspective it does not really matter whether genes are multiplied through the helper's own reproduction or that of relatives. The second explanation is known as *reciprocal altruism*; that is, helpful acts that are costly in the short run may produce long-term benefits if recipients return the favor. If I rescue a friend who almost drowns, and he rescues me under similar circumstances, both of us are better off than without mutual aid.

Wilson's *Sociobiology: The New Synthesis* summarized the new developments. It is an influential and impressive book predicting that all other behavioral sciences will one day see the light and convert to the creed of sociobiology. Confidence in this future was depicted in an amoebic drawing with pseudopods reaching out to devour other disciplines. Understandably, nonbiologists were piqued by what they saw as an arrogant attempt at annexation; but also within biology, Wilson's book provoked battles. Should Harvard be allowed to lay claim to an entire field? Some scientists preferred to be known as behavioral ecologists rather than sociobiologists, even though their theories were essentially the same. Moreover, like children ashamed of their old folks, sociobiologists were quick to categorize earlier studies of animal behavior as "classical ethology." That way everyone could be sure that ethology was dead and that we were onto something totally new.

Sociobiology represents a giant stride forward; it has forever changed the way biologists think about animal behavior. Precisely because of their power and elegance, however, the new theories have lured some scientists into a gross simplification of genetic effects. Behavior that at first sight does not conform to the framework is regarded as an oddity, even a mistake. This is best illustrated by a single branch of sociobiology, which has gotten so caught up in the Malthusian dog-eat-dog view of the world that it sees no room for moral behavior. Following Huxley, it regards morality as a counterforce, a rebellion against our brutish makeup, rather than as an integrated part of human nature.[9]

Calvinist Sociobiology

At the Yerkes Regional Primate Research Center, one chimpanzee has been named Atlanta and another Georgia. It is impossible for me to forget where I am, as I see both individuals on a daily basis. I moved to the Star of the South, as the city likes to call itself, to resume my study of the species that surpasses every other when it comes to similarity to our own. My tower office has a large window that overlooks the outdoor enclosure of twenty chimpanzees. The group is as close-knit as any family can be; they are together day and night, and several of the adults were born into the colony. One of these is Georgia, the rascal of the group. Robert Yerkes, a founder of primatology, once declared it "a securely established fact that the chimpanzee is not necessarily utterly selfish."[10] From everything I know about Georgia, she is not the sort of character Yerkes had in mind when he made that declaration six decades ago.

When we provision the colony with freshly cut branches and leaves from the forest around the field station, Georgia is often the first to grab one of the large bundles, and one of the last to share it with anybody else. Even her daughter, Kate, and younger sister, Rita, have trouble getting food. They may roll over the ground, screaming in a pitiful tantrum, but to no avail.

No, Yerkes must have thought of individuals such as Mai, an older high-ranking female, who shares quite readily not only with her children but also with nonrelatives, young and old. Or he may have thought of adult male chimpanzees, most of whom are remarkably generous when it comes to food distribution.

While a distinction between sharing and keeping means a lot in human society, it is sometimes lost in the language of a particular brand of sociobiology that takes the gene as absolute king. Gene-centric sociobiology has managed to reach a wide audience with its message that humans and other animals are entirely selfish. From this standpoint, the only difference between Mai and Georgia is in the *way* they pursue self-interest; whereas Georgia is just plain greedy, Mai shares food so as to make friends or receive return favors in the future. Both think only of themselves. In human terms, this interpretation amounts to the claim that Mother Teresa follows the same basic instinct as any inside trader or thief. A more cynical outlook is hard to come by.

Gene-centric sociobiology looks at survival and reproduction from the point of view of the gene, not the individual. A gene for bringing home food for one's children, for example, will ensure the survival of individuals likely to carry the same gene.[11] As a result, that gene will spread. Taken to its logical extreme, genes favor their own replication; a gene is successful if it produces a trait that in turn promotes the gene (sometimes summed up as "a chicken is an egg's way of making other eggs"). To describe such genetic self-promotion, Richard Dawkins introduced a psychological term in the title of his book, *The Selfish Gene*. Accordingly, what may be a generous act in common language, such as bringing home food, may be selfish from the gene's perspective. With time, the important addition "from the gene's perspective" was often forgotten and was eventually left out. All behavior was selfish, period.

Since genes have neither a self nor the emotions to make them selfish, one would think this phrase is just a metaphor. True, but when repeated often enough, metaphors tend to assume an aura of literal truth. Even though Dawkins cautioned against his own anthropomorphism of the gene, with the passage of time, carriers of selfish genes became selfish by association. Statements such as "we are born selfish" show how some sociobiologists have made the nonexistent emotions of genes into the archetype of true emotional nature. A critical article by Mary Midgley compared the sociobiologists' warnings against their own metaphor to the paternosters of the Mafiosi.

Pushed into a corner by a witty philosopher, Dawkins defended his metaphor by arguing that it was *not* a metaphor. He really meant that genes are selfish, and claimed the right to define selfishness any way he wanted. Still, he borrowed a term from one domain, redefined it in a very narrow sense, then applied it in another domain to which it is completely alien. Such a procedure would be acceptable if the two meanings were kept separate at all times; unfortunately, they merge to the extent that some authors of this genre now imply that if people occasionally think of themselves as unselfish, the poor souls must be deceiving themselves.

It is important to clear up this confusion, and to emphasize once and for all that the selfish gene metaphor says nothing, either directly or indirectly, about motivation, emotion, or intention. Elliott Sober, another philosopher interested in the semantic trappings of sociobiology, proposes a distinction between *vernacular egoism,* our everyday usage of the term, and *evolutionary egoism,* which deals exclusively with genetic self-promotion. A plant, for example, is able to further its genetic interests yet cannot possibly be selfish in the vernacular sense. A chimpanzee or person who shares food with others acts altruistically in the vernacular sense, yet we assume that the behavior came into existence because it served survival and reproduction, hence that it is self-serving in an evolutionary sense.[12]

There is almost no point in discussing the evolution of morality if we let the vernacular sense of our terminology be overshadowed by the evolutionary sense. Human moral judgment always looks for the intention behind behavior. If I lean out of a window on the fifth floor and unknowingly nudge a flowerpot, thereby killing a pedestrian on the sidewalk below, I might be judged awkward or irresponsible, but not murderous. The latter accusation would surely be heard, however, had someone watched me grab the pot and throw it at the person. The effect is the same, but the motives are absolutely crucial. Jury and judge would want to know which emotions I showed, the degree of planning involved, my relationship with the target, and so on. In short, they would want to fathom the psychology behind the act.

These distinctions are largely irrelevant within a sociobiology exclusively interested in the effects of behavior. In such a framework, no different values are attached to intended versus unintended results, self-serving versus other-serving behavior, what we say versus what we mean, or an honest versus a dishonest mistake. Having thus denied themselves the single most important handle on ethical issues, some sociobiologists have given up on explaining morality. William Hamilton, the discoverer of kin selection, has written that "the animal in our nature cannot be regarded as a fit custodian for the values of civilized man," and Dawkins urges us to cultivate pure, disinterested altruism because it does not come naturally. "We, alone on earth, can rebel against the tyranny of the selfish replicators."[13] By thus locating morality outside nature, these scientists have absolved themselves from trying to fit it into their evolutionary perspective.

An even more alarming position was adopted by George Williams in a commentary on Huxley's celebrated "Evolution and Ethics" lecture. Calling nature morally indifferent, as Huxley had done, was not enough for Williams, who preferred "gross immorality" and "moral subversiveness." He went on to demonstrate that "just about every . . . kind of sexual behavior that has been regarded as sinful or unethical can be found abundantly in nature." This conclusion was accompanied by a depressing enumeration of animal murder, rape, and wretchedness.[14]

Can we really pass judgment on other animals any more than we can on the flow of a river or the movement of nuclear particles? Does doing so get us beyond age-old stereotypes such as the hard-working bee, the noble horse, the cruel wolf, and the gluttonous pig? Granted, animals may possess standards of behavior, perhaps even ethical standards. Yet Williams was not measuring their behavior against their own standards, but against those of the culture of which he happens to be part. Since animals failed to meet his criteria, he declared nature, including human nature, our foe. Note, again, how vernacular egoism slips into a statement about the evolutionary process: "The enemy is indeed powerful and persistent, and we need all the help we can get in trying to overcome billions of years of selection for selfishness."[15]

By now, I am sure, the reader must have smelled the perfume Egoïste (an actual Chanel creation) to the point of either conviction or stupefaction. How in the world could a group of scientists come up with such a pale view of the natural universe, of the human race, of the people close to them, and of themselves (because we must assume that their theory knows no exceptions)? Do they not see that, to paraphrase Buddha, wherever there is shadow, there is light?

Underlying their position is a monumental confusion between process and outcome. Even if a diamond owes its beauty to millions of years of crushing pressure, we rarely think of this fact when admiring the gem. So why should we let the ruthlessness of natural selection distract from the wonders it has produced? Humans and other animals have been endowed with a capacity for genuine love, sympathy, and care—a fact that can and will one day be fully reconciled with the idea that genetic self-promotion drives the evolutionary process.

It is not hard to find the origin of the proposed abyss between morality and nature. The conviction is well established outside science. The image of humanity's innate depravity and its struggle to transcend that depravity is quintessentially Calvinist, going back to the doctrine of original sin. Tension between civic order and our bestial ancestry, furthermore, is the centerpiece of Sigmund Freud's *Civilization and Its Discontents*. Freud argues that we need to control and renounce our baser instincts before we can build a modern society. Hence, we are not dealing with a mere biological theory, but with a convergence between religious, psychoanalytical, and evolutionary thought, according to which human life is fundamentally dualistic. We soar somewhere between heaven and earth on a "good" wing—an acquired sense of ethics and justice—and a "bad" wing—a deeply rooted egoism. It is the age-old half-brute, half-angel view of humanity.

It must be rather unsatisfactory, to say the least, for gene-centric sociobiologists to be obliged to exclude one domain from their Theory of Everything. And not a trivial domain, but precisely the one many of us consider to be at the core of being human. Failure to account for morality in terms of genetic selfishness is the logical outcome of such reductionism. If we shrug off attempts to attribute love to hormones or hatred to brain waves—knowing that these attributions are only part of the story—it is good to realize that these are tiny jumps compared with the reduction of human psychology to gene action.

Fortunately, the current pendulum swing is away from such simplifications. It is toward attempts to explain living systems in their entirety, integrating many different levels. In the words of a recent task force of the National Science Foundation, "The biological sciences are moving away from the era of analytical reductionism . . . from taking biological systems apart to see what the pieces are and how they work, to putting the pieces back together to understand how the totality works together."[16]

One does not need to follow this holistic swing all the way to Gaia (the idea that the biosphere acts as a single organism) to agree that the current development indicates greater scientific maturity. In the New and Improved Sociobiology, animals still do everything to survive and reproduce, yet take their circumstances into account so as to choose the best course of action: from "survival machines" they have become "adaptive decision-makers." With so many degrees of freedom added, selfish-gene thinking can now safely be relegated to history as "classical sociobiology."

Have I been kicking a dead horse, then? I do not think so. Gene-centric sociobiology is the type best known to the general public. It is still widespread in certain academic circles, particularly those outside biology that have battled hard within their respective disciplines to stake out and defend an evolutionary approach. Furthermore, as a corollary to the belief in a natural world red in tooth and claw, there remains tremendous resistance, both inside and outside biology, to a terminology acknowledging beauty in the beast.

The sociobiological idiom is almost derisive in its characterization of animals. Given the image of biologists as nature buffs, it may be shocking for outsiders to learn that the current scientific literature routinely depicts animals as "suckers," "grudgers," and "cheaters" who act "spitefully," "greedily," and "murderously." There is really nothing lovable about them! If animals do show tolerance or altruism, these terms are often placed in quotation marks lest their author be judged hopelessly romantic or naive. To avoid an overload of quotation marks, positive inclinations tend to receive negative labels. Preferential treatment of kin, for instance, instead of being called "love for kin," is sometimes known as "nepotism."

As noted by economist Robert Frank (referring to a problem common to the behavioral sciences):

> The flint-eyed researcher fears no greater humiliation than to have called some action altruistic, only to have a more sophisticated colleague later demonstrate that it was self-serving. This fear surely helps account for the extraordinary volume of ink behavioral scientists have spent trying to unearth selfish motives for seemingly self-sacrificing acts.[17]

As a student of chimpanzee behavior, I myself have encountered resistance to the label "reconciliation" for friendly reunions between former adversaries. Actually, I should not have used the word "friendly" either, "affiliative" being the accepted euphemism. More than once I was asked whether the term "reconciliation" was not overly anthropomorphic. Whereas terms related to aggression, violence, and competition never posed the slightest problem, I was supposed to switch to dehumanized language as soon as the affectionate aftermath of a fight was the issue. A reconciliation sealed with a kiss became a "postconflict interaction involving mouth-to-mouth contact."

Barbara Smuts ran into the same resistance when she chose "friendship" as an obvious label for intimate relationships between adult male and female baboons. Can animals really have friends? was the question of colleagues who without blinking accepted that animals have rivals. Given this double standard, I predict that the word "bonding" will soon become taboo as well, even though it was initially coined by ethologists as a neutral reference to emotional attachment. Ironically, the term has since entered common English with precisely the meaning it tried to circumvent, as in "mother-child bond" and "male bonding." It is rapidly becoming too loaded for students of animal behavior.

Animals, particularly those close to us, show an enormous spectrum of emotions and different kinds of relationships. It is only fair to reflect this fact in a broad array of terms. If animals can have enemies they can have friends; if they can cheat they can be honest, and if they can be spiteful they can also be kind and altruistic. Semantic distinctions between animal and human behavior often obscure fundamental similarities; a discussion of morality will be pointless if we allow our language to be distorted by a denial of benign motives and emotions in animals.

An intriguing expression of emotion occurred once when, in the middle of the day, our entire chimpanzee colony unexpectedly gathered around Mai. All the apes were silent, staring closely at Mai's behind, some of them carefully poking a finger at it and then smelling their finger. Mai was standing half upright, with her legs slightly apart, holding one hand between her legs. Remarkably, an attentive older female mimicked Mai by cupping her hand between her own legs in exactly the same fashion.

After approximately ten minutes, Mai tensed, squatted more deeply, and passed a baby, catching it in both hands. The crowd stirred, and Atlanta, Mai's best friend, emerged with a scream, looking around and embracing a couple of other chimpanzees next to her, one of whom uttered a shrill bark. Mai then went to a corner to clean the baby and consumed the afterbirth with gusto. The next day Atlanta defended Mai fiercely in a fight, and during the following weeks she frequently groomed Mai, staring at and gently touching Mai's healthy new son.

This was the very first time I witnessed a chimpanzee birth. I have seen several macaque births, though, and the big difference is that other macaques do not approach the mother. It is hard to tell if they are even interested; there is no obvious excitement or curiosity about the delivery. Positive interest occurs only after the amniotic sac has been removed and the infant has been cleaned. For macaques are extremely attracted to newborns. Our chimpanzees responded much earlier; they seemed as much taken with the process as with its outcome. It is entirely possible that the emotional reaction of Atlanta (who has had quite a few infants of her own) reflected *empathy*, that is, identification with and understanding of what was happening to her friend.[18]

Needless to say, empathy and sympathy are pillars of human morality.

A Broader View

A climbing orangutan grasps a branch with one hand, holding on tightly until the other hand has found the next branch. Then the roles are reversed, and the first hand releases its grip in order to get hold of another branch. Elias Canetti, in *Crowds and Power,* noticed a connection between the ancient arboreal function of one of our most versatile organs and the universal human ritual of barter and trade: climbing through the trees may have predisposed us for economic exchange, since both activities depend on the careful coordination of grasp and release. With his goods held out in one hand, the tradesman reaches with the other for his partner's goods, mindful not to release anything before his grip on the desired goods is secure. Failure to perform this sequence in the right order or with the right timing may have fatal consequences in the trees in the same way that it may leave the human trader empty-handed. Material exchange has become second nature to us; most of the time we reflect as little on the risks as does a monkey racing through the canopy.

Canetti's is a fascinating parallel, yet there exists of course no causal connection. Otherwise the octopus would be the champion merchant of the animal kingdom, and animals without hands, such as dolphins and bats, would be excluded as possible traders. It is precisely bats, mammals with front limbs transformed into wings, who provide us with some of the first evidence for give-and-take relations in animals. Gruesome as it may sound, vampire bats trade meals by regurgitating blood to one another. At night these bats stealthily lap blood from a small patch of flesh exposed by razor-sharp teeth on a sleeping mammal, such as a horse or cow. With their bellies full, the bats return to the hollow tree in which they spend the day. We know about their blood economy because the bats sometimes share their roost with a scientist who spends hours on his back, legs outside and torso inside an opening at the base of the tree, peering upward to collect behavioral information along with the inevitable bat droppings.

Having tagged his subjects with reflective bands in order to recognize them in the dark, Gerald Wilkinson noticed that mother bats often regurgitate blood to their offspring. While this is not too surprising, the investigator saw other combinations share on twenty-one occasions—mostly individuals who often associated and groomed. There appeared to be a "buddy system" of food exchange, in which two individuals could reverse roles from night to night, depending on how successful each had been in finding blood. Because they are unable to make it through more than two nights in a row without food, it is a matter of life or death for vampire bats to have such buddies. Although the evidence is still meager, Wilkinson believes that these animals enter into social contracts in which each occasionally contributes part of a meal so as to be able to solicit a life-saving return favor during less favorable times.

Petr Kropotkin would have loved these little bats, as they exemplify the evolutionary principle advocated in his famous book *Mutual Aid,* which was first published in 1902. Though bearded and an anarchist, Kropotkin must not be thought of as a wild-eyed zealot. Stephen Jay Gould assures us, "Kropotkin is no crackpot."[19] Born a Russian prince, and very well educated, he was a naturalist and intellectual of great distinction. He was offered the position of secretary to the Imperial Geographical Society in Saint Petersburg, and later, during his exile in England, a chair in geology at Cambridge University. He declined both positions inasmuch as they would have interfered with his political activities, which aimed, according to a comrade, at opposing with an ecstasy of expiation the very injustice of which fate had made him the involuntary beneficiary.

Animals, Kropotkin argued in *Mutual Aid,* need to assist one another in their struggle for existence; a struggle, not so much of each against all, but of masses of organisms against the adversity of their environment. Cooperation is common, as when beavers together dam off a river or when horses form a protective ring against attacking wolves. Kropotkin did not stand alone in his emphasis on sociality and communion among animals: an entire generation of Russian scientists was uncomfortable with the primacy given in evolutionary thought to competition. Daniel Todes, in a fascinating treatise on Russian natural science aptly entitled *Darwin without Malthus,* argued that there may have been geographical reasons for this different outlook.

Whereas Darwin found inspiration in a voyage to rich, abundant tropical regions, Kropotkin at the age of nineteen set out to explore Siberia. Their ideas reflect the contrast between a world where life is easy, resulting in high population densities and intense competition, and one where life is harsh and filled with unpredictable dangers. When discussing evolution, Kropotkin and his compatriots always had their sparsely populated continent in mind, with its rapidly changing weather and extreme seasonality. He described climatic calamities that could render a territory as large as France and Germany combined absolutely impracticable for ruminants, in which horses could be scattered by the wind, and entire herds of cattle could perish under piles of snow.

[These calamities] made me realize at an early date the overwhelming importance in nature of what Darwin described as "the natural checks to over-multiplication," in comparison to the struggle between individuals of the same species for the means of subsistence, which may go on here and there, to some limited extent, but never attains the importance of the former. Paucity of life, under-population—not over-population—being the distinctive feature of that immense part of the globe which we name Northern Asia, I conceived since then serious doubts as to the reality of that fearful competition for food and life within each species, which was an article of faith with most Darwinists, and, consequently, as to the dominant part which this sort of competition was supposed to play in the evolution of new species.

Kropotkin objected vehemently to the depiction of life as a "continuous free fight" and a "gladiator's show" made popular by the same Huxley who five years later, just before his death, partially reversed and softened his position to introduce morality as humanity's saving grace. Playing down Huxley's competitive principle, Kropotkin instead saw a communal principle at work: cooperation and mutual assistance among animals arose in response to the common enemy. The idea of a common enemy is perhaps the most significant of all of Kropotkin's ideas. In his mind it referred to the hostile environment in which many animals try to exist and multiply.

Kropotkin's analysis had serious flaws, and he sprinkled *Mutual Aid* with highly selective, often dubious examples to make his case. He had a (not so) hidden revolutionary agenda, and read political preferences into nature to the point that he totally overlooked its nasty side. He stated that "in the face of free Nature, the unsociable instincts have no opportunity to develop, and the general result is peace and harmony." Kropotkin, however, was writing in direct response to people who reduced everything in nature to savage, unmitigated combat.[20] Their position too could hardly be considered free from ideological bias. Russian scientists of that period saw the gladiatorial view as a concoction of the British upper class to defend the status quo.

Kropotkin cast his arguments in terms of survival of the group, or the species as a whole. Rejection of this view, known as group selection, marked the rise of sociobiology. Contemporary biologists in general do not believe that behavior evolved for a greater good. They assume that if bats, bees, dolphins, and other animals help one another, there must be benefits for each and every participant or their kin, otherwise the trait would not have spread.[21]

Old ideas never die completely, and group selection has been staging a gradual comeback.[22] It is also good to realize that Kropotkin was in excellent company in his belief that the success of the group matters: Darwin himself leaned toward group selection when tackling the issue of morality. He literally saw one tribe gain advantage over another:

> At all times throughout the world tribes have supplanted other tribes; and as morality is one element in their success, the standard of morality and the number of well-endowed men will thus everywhere tend to rise and increase.[23]

I should not leave the impression that Darwin and Kropotkin were in the same league as thinkers about evolution. Darwin argued his case much more systematically and coherently, and with vastly greater knowledge, than did the Russian naturalist. *Mutual Aid* was no match for Darwin's powerful exposé of the principles of natural selection, and Kropotkin, despite profound disagreements with Darwin's followers, never wavered in his admiration for the master himself.

Mutual aid became a standard ingredient of sociobiology, not through the writings of Kropotkin but because of a single article that presented the concept with such precision and clarity that modern biologists could not ignore it. I still remember the excitement when in 1972, with a number of students at the University of Utrecht, I analyzed Robert Trivers' "Evolution of Reciprocal Altruism." It remains one of my favorite articles. Rather than simplifying the relation between genes and behavior, it pays full attention to the intermediate levels such as emotions and psychological processes. It also distinguishes different types of cooperation based on what each participant puts into and gets out of it. For example, cooperation with immediate rewards does not qualify as reciprocal altruism. If wild dogs together bring down a wildebeest, all hunters benefit at the same time. Similarly, if a dozen pelicans form a semicircle in a shallow lake to herd small fish with their paddling feet, all birds profit when they together scoop up the fish. Because of the instant payoff, this kind of cooperation is widespread.

Reciprocal altruism, on the other hand, costs something before it delivers benefits. It has the following three characteristics:

1. The exchanged acts, while beneficial to the recipient, are costly to the performer.
2. There is a time lag between giving and receiving.
3. Giving is contingent on receiving.

This process is evidently a lot more complicated than simultaneous cooperation. There is, for example, the problem of the first helpful act—a gamble, since not every partner necessarily follows the rules. If I help you move your piano, I cannot be sure that you will do the same for me in the future. Or if one bat shares blood with another, there is no guarantee that the other will return the favor the next day. Reciprocal altruism differs from other patterns of cooperation in that it is fraught with risk, depends on trust, and requires that individuals whose contributions fall short be shunned or punished, lest the whole system collapse.

Reciprocal altruism does not work for individuals who rarely meet or who have trouble keeping track of who has done what for whom: it requires good memories and stable relationships, such as are found in the primates. Monkeys and apes make sharp distinctions between kin and nonkin as well as between enemies and friends. The chief purpose of friendship being mutual support, it is only natural that

Bonobo

Chimpanzee

Baboon

Stump-tailed macaque

Rhesus macaque

Capuchin

The order Primates includes nearly two hundred different species, ranging in size from the 100-gram pygmy marmoset to the 100-kilogram male gorilla. The six nonhuman primates most often referred to in the text are portrayed here. Chimpanzees and bonobos, as apes, belong to the same hominoid branch of the evolutionary tree as our own species; the other four primates depicted here are monkeys. The capuchin is the single New World monkey.

CLOSENESS

Mammalian societies often are intimate groups in which everyone knows everyone else. Birth and death mark entry to and exit from the system: newborns are the center of attention, and attachments linger long after an individual's demise. The relationship between mother and offspring provides a blueprint for all other affiliations. Handicapped individuals may be fully accepted in the group.

A free-ranging Japanese monkey in Jigokudani Park born without hands and feet. Mozu (portrait on next page) has survived the harsh climate of this mountainous region for two decades.

At the age of three months Azalea, a retarded rhesus female born to an aging mother, showed unusual facial features. At the age of two years—when her peers were already quite independent and enterprising—Azalea was still being carried around and held by older kin—as below by a sister. *(Wisconsin Primate Center)*

A rhesus monkey with cuddly baby features. *(Wisconsin Primate Center)*

A snow-white newborn in its mother's lap, surrounded by other stump-tailed macaques. *(Wisconsin Primate Center)*

The capuchin monkey is sometimes referred to as the South American chimpanzee, even though capuchins and chimpanzees are only distant relatives. The two species share traits such as tool use, large brains, an omnivorous diet, slow development, and long lifespan. According to recent studies, food sharing needs to be added to this list. The photograph shows a mother and infant of the common brown, or tufted, capuchin. *(Yerkes Primate Center)*

Millions of years of selection for mothers who pay close attention to the needs of their offspring have promoted strong nurturing tendencies in mammalian females. A bonobo with her daughter. *(San Diego Zoo)*

Weaning compromises are sometimes worked out between mother and off-spring. One juvenile chimpanzee has developed a habit of "nursing" under her mother's armpit; another, of sucking on his mother's lower lip. *(Yerkes Field Station)*

Chimpanzees have many ways to attract sympathy. An adolescent female persuades her mother to go somewhere. First she whimpers in a plaintive voice, stretching out both hands (facing page). The daughter's face changes when her mother approaches (above, top). The two end up traveling together. *(Yerkes Field Station)*

What is remarkable about these sleeping rhesus monkeys is that they are the alpha female (left) and the lowest-ranking matriarch. Most of the time, hierarchical distance prevents such association. *(Wisconsin Primate Center)*

Rather than treasuring individual autonomy, our primate relatives devote a great deal of time to intimate relationships: a female Japanese monkey grooms a male. *(Jigokudani Park, Japan)*

Fifteen months after having lost her mother, Agatha still regularly returns to the fatal spot to gently turn and feel her mother's skull. *(From* Elephant Memories *by Cynthia Moss; Amboseli National Park, Kenya)*

Robert Trivers' theory of reciprocal altruism, published in 1971, is and will remain the centerpiece of any viable theory of moral evolution. On June 20, 1992, I interviewed Trivers about the history of his ideas at a meeting of the Gruter Institute for Law and Behavioral Research, in Squaw Valley, California.

Q: How did you come to write about reciprocal altruism?

A: I became aware of Hamilton's paper in 1966 or 1967, which explained altruism among kin. It immediately raised the issue of what other kinds of altruism there might be. Friendship is, I think, as strong an emotion as kinship ties . . . especially in the sort of fractured society that we live in. Naturally, I thought of "you scratch my back—I scratch yours," which is the folk expression for reciprocity.

Q: You did not have a lot of animal examples?

A: No, I knew about alarm calls in birds and cleaning symbiosis in fish, but none of the animal examples was terribly convincing. I thought coalitions in baboons might be a good candidate, but nobody really knew if they were reciprocal. Then I had a lucky break. I happened to take a course on morality, but the guy who gave it was so confused about the issue that I became more interested in his graduate student, Dennis Krebs, who was writing about what psychologists call "prosocial behavior" [he pulled a disgusted face]. Krebs's paper, which was later published, contained plenty of good examples of human altruism with absolutely no reference to their function or evolution. So all I had to do was reorganize this information along the lines of my ideas.

I was fond of arguing from humans to other creatures, anyway—rather than the opposite—perhaps because I came to biology rather late in life; I did not take a biology course until I was twenty-four.

Q: Reading between the lines, I recognize in your paper the same sort of social commitment that led Kropotkin to develop his ideas. . . .

A: People keep asking about Kropotkin. You know, I have never read the anthropologists who wrote about reciprocity, and I have never read Kropotkin. [He picked up a tiny ant that walked on the table, and explained how to tell that it was a male.]

But you're right about my political preferences. When I left mathematics and cast about what I was going to do in college, I said [ironic tone], "All right, I'll become a lawyer and fight for civil rights and against poverty!" Someone suggested that I take up U.S. history, but you know at that time, in the early 1960s, their books were entirely self-congratulatory. I ended up in biology.

Because I remained a political liberal, for me, emotionally, to see that just pursuing this scratch-my-back argument would generate rather quickly a reason for justice and fairness was very gratifying, because it was on the other side of the fence of that awful tradition in biology of the right of the strongest.

such relationships develop primarily between individuals with common interests. I know two inseparable females, Ropey and Beatle, in a large group of rhesus macaques that I followed for a decade at the Wisconsin Regional Primate Research Center. They are approximately the same age, and Beatle's family ranks just below Ropey's in the matrilineal hierarchy. I would have sworn they are sisters: they do everything together, often grooming each other, and softly lipsmacking at each other's infants. According to the center's records, however, Ropey and Beatle are totally unrelated.

This friendship typifies a general rule, called the *similarity principle,* which probably has much to do with reciprocal exchange. According to our computer records on hundreds of female-female relationships, female monkeys select friends on the basis of closeness in age and rank. In a macaque society, with its strict hierarchy, females of similar status experience harassment by the same dominants, and need to keep the same subordinates in line. Between females of similar age, important life stages are synchronized: they play together when they are young; they deliver their first offspring at about the same time; their daughters and sons grow up together, as do their grandchildren. With so much in common, it is no wonder that these females gravitate toward each other rather than toward females of different status, or much older or younger in age. The principle is also illustrated by Mozu's successful attempt to connect with females her own age in the dominant subdivision after fissioning of the Jigokudani group.

This is not to say that exceptions do not occur. I still recall the day on which the alpha female of our rhesus group walked up to the lowest-ranking matriarch, presented herself for grooming, and after a long session fell asleep in the midst of this female's family. Alpha was completely out of place as far as status was concerned, with her head serenely resting on the back of the other matriarch while this female's offspring, after much hesitation, huddled against the two in a sleeping cluster that superficially looked like any other.

Such special moments only serve to confirm the rule, which holds for our species as well. Besides age and socioeconomic status, the similarity principle in humans includes political preference, religion, ethnic background, IQ level, education, physical attractiveness, and height. Birds of a feather tend to like and love each other, to the extent that researchers can statistically predict if dating couples will break up or stay together based on how well they match on various dimensions. As in rhesus monkeys, these matching rules probably relate to prospects for cooperation: the more traits and interests one shares with another person, the easier it will be to get along, and the broader the basis for a give-and-take relationship.

That primates, consciously or unconsciously, seek the company of individuals with whom profitable partnerships are possible perhaps sounds too obvious to deserve attention. Not, however, if one considers the tradition in evolutionary biology of pitting the interests of one individual against those of another. Kropotkin's intuitive opposition to an exclusive focus on competition, and Trivers' concrete alternative have yet to sink in fully. Lack of appreciation of the widespread conflux of interests among members of the same species has led respectable biologists to ask the wrong questions. Puzzled by the fact that animals almost never fight to the death, some have felt the need to demonstrate mathematically that there are drawbacks to all-out competition. In doing so, they have zoomed in on the physical risks rather than the social consequences of fighting. The possibility that animals might wage "limited wars" because they know each other, need each other, hence value good relationships, never entered into the equations.

There can be little doubt that in many species the strong can annihilate the weak. In a world of mutual dependency, however, such a move would not be very smart. The real problem is not why aggression is tempered—it needs to be—but how cooperation and competition coexist. How do individuals strike a balance between serving their own interests and operating as a team? How are conflicts resolved without damage to social ties? If these questions (which follow directly from mutual aid as a factor in society) sound familiar, it is because we face them every day at home and at work.

The Invisible Grasping Organ

As far back as 1714 a Dutch philosopher, Bernard de Mandeville, achieved international fame by exploiting the same shock effect that has since served so many popular authors. In a lengthy poem, *The Fable of the Bees: or, Private Vices, Public Benefits,* he ascribed the loftiest aspects of human life to our vilest qualities, comparing civilization to a beehive in which all individuals happily gratify each other's pride and vanity:

> Thus every part was full of vice
> Yet the whole mass a paradise.[24]

Whereas de Mandeville provided a mere satire, the idea of public blessings derived from the pursuit of self-interest gained respectability when the father of economics, Adam Smith, pronounced self-interest society's guiding principle. In a passage from the *Wealth of Nations,* first published in 1776, Smith saw each individual as being "led by an invisible hand to promote an end which was no part of his intention. Nor is it always the worse for society that it was no part of it. By pursuing his own interest he frequently promotes that of the society more effectually than when he really intends to promote it."[25]

The crux of Smith's famous invisible hand metaphor is that of a gap between intention and consequence; our actions may mean something entirely different in the bigger scheme of things than they do to us individually. Thus, life in the city depends on the professional services of bakers, mechanics, and storekeepers, whereas these people are only carving out a living for themselves and unknowingly are driven to serve the larger whole. To this day, the metaphor is popular with economists: a recent *New Yorker* cartoon showed a gathering of economists kneeling in the grass, hoping for a huge invisible hand to appear in the sky.

Smith's views were complex, however. As a moral philosopher, he knew full well that it would be hard to hold a society together purely on the basis of egoism. Like Huxley, Smith mellowed with age; he consecrated the final years of his life to a revision of *A Theory of Moral Sentiments,* expanding on his earlier belief in unselfish motives. Throughout this work Smith rejected the self-love thesis of de Mandeville, observing in the very first sentence that man possesses capacities "which interest him in the fortune of others, and render their happiness necessary to him, though he derives nothing from it, except the pleasure of seeing it."[26] This passage still stands as one of the most succinct and elegant definitions of human sympathy, a tendency Smith believed present in even the staunchest ruffian.

What makes the invisible hand metaphor so powerful is the idea of simultaneous micro and macro realities: the reality in the mind of each individual is not the same as the reality that emerges when many individuals interact. At one level we do A for reason B, while at another level A serves purpose C. Biologists are familiar with such multilevel thinking. For example, sex serves reproduction, yet animals engage in it without the slightest notion of its function; they are not driven by any desire to reproduce, only by sexual urges (as are humans most of the time). Similarly, members of a species do not need to have the benefits of mutual aid in mind when they help one another; these benefits may be so indirect and delayed in time that they matter only on an evolutionary timescale.

Imagine that you and I are sitting in individual boats adrift in the large swimming pool of an immense cruise ship. The ship has a slow and steady course to the north, but we care only about the direction of our little boats and cannot see beyond the body of water within which we are maneuvering. Even though I decide to head west and you to head south, to the outside observer both of us wind up going north. Since our experiences do not match our eventual destination, we live in separate proximate and ultimate realities.[27]

Intention and consequence need not be independent, however, particularly in our species. Often we have a reasonable understanding of the effects of our actions, especially when these effects are immediately obvious. Thus, it cannot escape us that one function of cooperative behavior is what appears to be its antipode: competition. Is not cooperative competition what team sport and party politics are all about?

In the primate order, the most widespread and best-developed collaboration is alliance formation, defined as two or more individuals banding together to defeat a third. For example, two male chimpanzees team up in order to overthrow the established ruler. The two challengers will swagger shoulder to shoulder in an intimidating manner, with their hair on end, often embracing or mounting each other directly in front of their rival, and of course supporting each other if it comes to an actual confrontation. By doing so for weeks or months on end, they wage a veritable war of nerves that may force the other male out of power. This is one of the most committed forms of animal cooperation that I know, one in which lives are literally at stake. Alpha males rarely go down without a fight.

Idealists such as Kropotkin tend to focus entirely on the agreeable side of cooperation, such as loyalty, trust, and camaraderie, while ignoring the competitive side. Although the Russian naturalist did refer to the role of a common enemy in fostering mutual aid, he conveniently ignored the possibility that the enemy might belong to one's own species. In *The Biology of Moral Systems* the biologist Richard Alexander presents our violent history of group against group and nation against nation as the ultimate reason why we attach so much value to the common good and to ethical conduct.

Alexander points out, however, that conflict between groups cannot be the whole explanation. Ants, for example, engage in terrible warfare on a massive scale, yet no one would argue that they have anything resembling a moral system. Thousands of them may be locked in deadly combat carried out in broad daylight on the sidewalks of our cities, while thousands more are recruited to join the massacre. Within each colony, however, harmony reigns. Ants form colonies of millions of individuals produced by and reproducing through a single female, the queen. With such overlapping reproductive interests, why should they compete with their own colony mates? And without conflicts of interest to be settled, what good would a moral system do?

The second condition for the evolution of morality, then, is conflict *within* the group. Moral systems are produced by tension between individual and collective interests, particularly when entire collectivities compete against one another.

If the need to get along and treat each other decently is indeed rooted in the need to stick together in the face of external threats, it would explain why one of Christianity's most heralded moral principles, the sanctity of life, is interpreted so flexibly, depending on which group, race, or nation the life belongs to. As recently as 1991, a war was declared "clean" and said to have been conducted with "clinical precision" despite the loss of more than a hundred thousand lives! Because the overwhelming majority of the dead in the Gulf War had fallen on the other side, Western media and politicians saw no need to burden our consciences.

Human history furnishes ample evidence that moral principles are oriented to one's own group, and only reluctantly (and never evenhandedly) applied to the outside world. Standing on the medieval walls of a European city, we can readily imagine how tightly life within the walls was regulated and organized, whereas outsiders were only important enough to be doused with boiling oil. There is of course great irony in Alexander's suggestion that the moral underpinnings of the community, on the one hand, and warfare and ethnic strife, on the other, are two sides of the very same coin. In modern times we highly value the first, yet feel embarrassed by the stubborn persistence of the second.

Both conditions for the evolution of morality apply to monkeys and apes. First, many species engage in intergroup conflict, mostly at low intensity but sometimes with extreme brutality. Wild male chimpanzees, for example, may take over a neighboring territory by systematically killing off the males of the other community. Second, while there is no lack of strife and competition within groups, we also know that primates have ways of resolving conflict nonaggressively. Because this was the subject of my previous book, *Peacemaking among Primates,* it is only logical that I investigate morality next. In doing so, I will take the perspective of the animals themselves in their day-to-day social life: their proximate reality. Insofar as the course of the big ship of evolution is concerned, I will rely on the combined ideas of Kropotkin, Trivers, and Alexander.

The one extension I would like to make is what I will call *community concern.* Inasmuch as every member benefits from a unified, cooperative group, one expects them to care about the society they live in and make an effort to improve and strengthen it, similar to the way the spider repairs her web and the beaver maintains the integrity of his dam. Continued infighting, particularly at the top of the hierarchy, may damage everyone's interests; hence the settlement of conflict is not just a matter of the parties involved, it concerns the community as a whole. I do not necessarily mean that animals make sacrifices for their community, but rather that each and every individual has a stake in the quality of the social environment on which its survival depends. In trying to improve this quality for their own purposes, they help many of their group mates at the same time. An example is arbitration and mediation in disputes; it is standard practice in human society—courts of law serve this function—but recognizable in other primates as well.

With their flamboyant orange coats of thick, long hair and their gentle blue faces, golden monkeys are perhaps the most gorgeous primates in the world. In the wild, these rare langurs from China occur in aggregations of up to three hundred or four hundred individuals. Field-workers believe that within these groupings exist many smaller units of a single adult male, several females, and their dependent offspring. As in other primates with one-male units, male golden monkeys are twice the size of females. Even if the male therefore is absolute king, the integrity of his unit probably depends as much on how well the females get along as on his dominance over them.

RenMei Ren, a primatologist at the University of Peking, found that male golden monkeys actively promote peaceful coexistence among their females, intervening in virtually every female altercation. They either break up the fight by dispersing the combatants or prevent further hostilities by taking up a position between them. Occasionally the male will calm tempers by gesturing to both female rivals, turning from one to the other with a friendly facial expression or combing his fingers through the long hair on each female's back. Once, when a male was removed from a captive group for health reasons, Ren noticed excessive violence among the females. The situation improved immediately upon the male's return.

Mediation "from above" by a high-ranking individual is obviously easier than mediation "from below." The principle remains that of a third party promoting good relationships, yet the risks are greater in the latter case because both combatants may take out their tensions on a low-ranking mediator. Nevertheless, such mediation does occur—but as far as we know, only in the chimpanzee. When male chimpanzees fail to reconcile after a confrontation, they sometimes sit a couple of meters apart as if waiting for their opponent to make the first move. The uneasiness between them is obvious from the way they look in all directions—the sky, the grass, their own body—while scrupulously avoiding eye contact with each other. On a previous occasion I described how this deadlock may be broken by a female mediator:

> Especially after serious conflicts between two adult males, the two opponents sometimes were brought together by an adult female. The female approached one of the males, kissed or touched him or presented towards him and then slowly walked towards the other male. If the male followed, he did so very close behind her (often inspecting her genitals) and without looking at the other male. On a few occasions the female looked behind at her follower, and sometimes returned to a male that stayed behind to pull at his arm to make him follow. When the female sat down close to the other male, both males started to groom her and they simply continued grooming after she went off. The only difference being that they groomed each other after this moment, and panted, spluttered, and smacked more frequently and loudly than before the female's departure.[28]

During our studies of the world's largest chimpanzee colony at the Arnhem Zoo in the Netherlands, such mediating behavior was repeatedly observed. It allows male rivals to approach each other without taking any initiative, without making eye contact, and perhaps without losing face. Females in this colony also approached males who were gearing up for a confrontation (male chimpanzees may sit for five or ten minutes with their hair erect, swaying from side to side and hooting, before actually turning against each other) to gently pry open their hands and "confiscate" weapons such as heavy sticks or rocks. If female chimpanzees appear concerned about male relationships, they have excellent reason: males have a tendency to take tensions out on females.

Community concern, then, is expressed in the amelioration of social relationships among others to the intervener's own advantage. This represents a first step toward a system such as human morality that actually elevates community interests above individual interests. Not that the two ever become disconnected—natural selection would never produce such an arrangement—yet the focus may gradually shift away from the individual toward the collective, or rather from egocentric to shared community concerns. Insofar as certain qualities of the social environment benefit large numbers of individuals, it is logical for the participants to encourage each other in shaping the community along these lines. The more developed the system of encouragement, the more important common goals become relative to private ones. Some of this encouragement may take the form of what Alexander has termed *indirect reciprocity*: rather than through a direct exchange of favors between two individuals, as in reciprocal altruism, helpful behavior may pay off via third parties.

Imagine that you have put your life at risk to save little John, who was playing on the railroad tracks. Within hours, the entire village will know what happened because people take careful note of social events around them. Your standing as a fine and trustworthy person will immediately go up one or two notches, which may help your contacts and your business. It is not little John who is doing you a return favor, but the community as a whole. It rewards behavior that improves the quality of life. If all community members keep an eye on how each one of them responds to others in need, they will quickly learn who is likely to help and who is not. Once doing good is appreciated at the group level, it does not need to be rewarded on a tit-for-tat basis to yield benefits.

In a moral community, it matters not just what I do to you and what you do to me, but also what others think of our actions. Perceptions become a major issue. For this reason Adam Smith introduced an imaginary *impartial spectator* capable of evaluating social events with sympathy and understanding. Our actions are mirrored in the eyes of the spectator in the same way that everything we do is reflected in the responses of our group. Theories of moral evolution need to assign a significant role to such outside attention, hence concern themselves with the community level. Although Darwinian at their core, they thus begin to transcend the exclusive focus on the individual by taking as their subject the way conflicts of interest are

Conditions for the evolution of morality:

1. **Group value** — Dependence on the group for finding food or defense against enemies and predators
2. **Mutual aid** — Cooperation and reciprocal exchange within the group
3. **Internal conflict** — Individual members have disparate interests

Given the above conditions, intragroup conflict needs to be resolved through a balancing of individual and collective interests, which can be done at both dyadic and higher levels:

1. **Dyadic level** — One-on-one interaction between individuals, such as direct reciprocation of aid and reconciliation following fights
2. **Higher levels** — Community concern, or care about good relationships between others, expressed in mediated reconciliation, peaceful arbitration of disputes, appreciation of altruistic behavior on a groupwide basis (indirect reciprocity), and encouragement of contributions to the quality of the social environment. (The last two may be limited to human moral systems; the first two are more widespread.)

resolved and societies are constructed. If each individual tries to shape its social environment and receive feedback about how these efforts affect others, society becomes essentially an arena of negotiation and give-and-take.

Our thinking here begins to resemble a *social contract,* getting close to philosophical, psychological, sociological, and anthropological theories of human sociality. While some may regard this as a watering down of the evolutionary approach, it is an inevitable development. Faced with a mountain as intimidating as morality, we either acquire the theoretical gear to climb all the way to the top, or we stay in the foothills with a few simplistic notions.[29]

Ethology and Ethics

In the 1940s a special label became necessary to distinguish the study of animal behavior in nature from the laboratory experiments of behaviorists on white rats and other domesticated animals. The chosen name was *ethology,* and its most famous representative became the Austrian zoologist Konrad Lorenz. The image of Lorenz being followed by a cohort of honking geese, or calling his tame raven out of the sky, was quite different from that of B. F. Skinner with his hand around the wings of a pigeon, placing the bird in a so-called Skinner box. The distinction was not merely in the personal relation with the subject of study and the way behavioral information was being extracted, but also in the favored explanation—with one school emphasizing instinct, the other learning.

The term "ethology" comes from the Greek *ethos,* which means character, both in the sense of what is characteristic of a person or animal and in the sense of moral qualities. Thus, in seventeenth-century English an ethologist was an actor who portrayed human characters on stage, and in the nineteenth century ethology referred to the science of building character. While it cannot be denied that there are some real characters among contemporary ethologists, the meaning of the word changed when a Frenchman, Isidore Geoffroy-Saint-Hillaire, selected it in 1859 to denote the study of animal behavior in the natural habitat. The term survived in a small circle of French biologists until, almost a century later, it became popular among other continental students of animal behavior. When ethology later reached Great Britain, the term gained its current status in Webster's as "the scientific study of the characteristic behavior patterns of animals" (although most ethologists would probably change "scientific" to "naturalistic").[30]

Early ethology stressed instinct—suggesting purely inborn behavior—yet was by no means blind to other influences. In fact, one of its strongest contributions was the study of imprinting, a learning process. Ducklings and goslings are not born with any detailed knowledge of their species; they acquire information in the first hours of their life. Normally, they do so by watching and following their mother, but they may take cues from any moving object encountered during the sensitive period. In nature, the chance of becoming attached to the wrong kind of object is minimal, but scientists have been able to make birds follow toy trucks and bearded zoologists. Therefore, what is innate in these birds is not so much any precise knowledge of their species, but a tendency to acquire this knowledge at a critical stage in life.

The mind does not start out as a tabula rasa, but rather as a checklist with spaces allotted to particular types of incoming information. Predispositions to learn specific things at specific ages are widespread, the best human example being language acquisition. We are not born with a particular language, but with the capacity to organize rather chaotic information into an orderly linguistic structure. Before the age of seven, we are so incredibly successful at this that our mind seems like a sponge, ready to absorb all sorts of subtleties of the language spoken around us. Whatever we learn later on never compares to the ease of use of our native tongue. I speak from experience: both at home and on the job, I have used two nonnative languages on a daily basis for decades, yet every sentence in these languages still requires a fraction of a second more processing time than my native Dutch. Secondary languages are donned like clothes; only the native tongue feels like skin.[31]

Human morality shares with language that it is far too complex to be learned through trial and error, and far too variable to be genetically programmed. Some cultures permit the killing of newborns, whereas others debate abortion of the unborn. Some cultures disapprove of premarital sex, whereas others encourage it as part of a healthy sexual education. The gravest error of biologists speculating about the origin of morality has been to ignore this variability, and to downplay the learned character of ethical principles.

Possibly we are born, not with any specific social norms, but with a learning agenda that tells us which information to imbibe and how to organize it. We could then figure out, understand, and eventually internalize the moral fabric of our native society. Because a similar learning agenda seems to regulate language acquisition, I will speak of *moral ability* as a parallel to language ability. In a sense, we are imprinted upon a particular moral system through a process that, though hundreds of times more complicated than the imprinting of birds, may be just as effective and lasting. And, as in birds, the outcome may deviate from the norm. A friend of mine attributes the thrill she gets from smuggling to the praise she received as a young girl whenever she managed to stash away food (during World War II she spent years in a Japanese concentration camp). The criminal justice system of course encounters far more serious deviations, and these too can often be traced to lessons received, or not received, during sensitive phases of moral development.[32]

Does this make morality a biological or a cultural phenomenon? There really is no simple answer to this sort of question, which has been compared to asking whether percussive sounds are produced by drummers or by drums. If we have learned anything from the debate

IM, the two-headed snake. (Courtesy of Gordon Burghardt; University of Tennessee)

between ethologists and behaviorists, it is that nature and nurture can be only partially disentangled. The same is true of relatively simple processes, such as the effect of light on plants. If a plant in a sunny spot grows taller than one in the shade, it is not because of either genetics or the environment, but both. True, the size difference is produced by variable light conditions, but it is equally true that light matters the way it does because of the genetic makeup of this particular organism; other plants thrive in the shade and shrivel in the sun. Environmental influences—including human culture—vary with the genetic "substrate" on which they act.

Perhaps a snake named IM is the best analogue to the false dichotomies of the past. This double-headed monstrosity, which I once held in my hands (after having been reassured about its friendly character), lives at the University of Tennessee. Gordon Burghardt, an American ethological psychologist, explained that the left head had been named "Instinct" and the right head "Mind" because of the perennial conflict between these concepts. The two heads of the snake literally fight over prey, each attempting to swallow the mouse or rat and thereby lengthening the ingestion process from minutes to hours. It is a futile battle, since the food ends up nourishing exactly the same body.

Nevertheless, scientists retain a tendency to claim the primacy of one head or the other, and ethologists have been no exception. Virtually every existing moral principle has by now been biologically explained, a dubious genre of literature going back to Ernest Seton's *Natural History of the Ten Commandments,* published in 1907. Other biblical titles have followed, principally in the German language, spelling out how moral principles contribute to survival of the species.[33] If law and religion prohibit the killing of fellow humans, the reasoning goes, it is in order to prevent extinction of the human race. Supported by the then-prevailing opinion that no animal ever lethally attacks a member of its own species, this argument sounded logical enough. But we know now that we are by no means the only murderous species, not even the only "killer ape." In the Arnhem chimpanzee colony, for example, one male was killed and castrated by two others in a fight over sex and power. The steadily growing list of species in which lethal aggression occurs—even if rarely—illustrates the weakness of species-survival arguments.[34]

Much of this literature assumes that the world is waiting for biologists to point out what is Normal and Natural, hence worth being adopted as ideal. Attempts to derive ethical norms from nature are highly problematic, however. Biologists may tell us how things are, perhaps even analyze human nature in intricate detail, yet there is no logical connection between the typical form and frequency of a behavior (a statistical measure of what is "normal") and the value we attach to it (a moral decision). Lorenz came close to confusing the two when he was disappointed that the perfect goose marriage, with the partners faithful unto death, was actually quite rare. But perhaps Lorenz was only titillating his readers with his favorite birds' "shortcomings," because he also gave us his student's wonderful retort: "What do you expect? After all, geese are only human!"[35]

Known as the *naturalistic fallacy,* the problem of deriving norms from nature is very old indeed. It has to do with the impossibility of translating "is" language (how things are) into "ought" language (how things ought to be). In 1739 the philosopher David Hume made these points in *A Treatise of Human Nature:*

> In every system of morality, which I have hitherto met with, I have always remarked that the author proceeds for some time in the ordinary way of reasoning, and establishes the being of God, or makes observations concerning human affairs; when of a sudden I am surprised to find, that instead of the usual copula-

tion of propositions, *is*, and *is not*, I meet with no proposition that is not connected with an *ought*, or an *ought not*. This change is imperceptible; but is, however, of the last consequence. For as this *ought*, or *ought not*, expresses some new relation or affirmation, it is necessary that it should be observed and explained; and at the same time that a reason should be given, for what seems altogether inconceivable, how this new relation can be a deduction from others, which are entirely different from it.[36]

To put the issue of ethics back into ethology in a more successful manner, we need to take note of the chorus of protest against previous attempts. Philosophers tell us that there is an element of rational choice in human morality, psychologists say that there is a learning component, and anthropologists argue that there are few if any universal rules. The distinction between right and wrong is made by people on the basis of how they would like their society to function. It arises from interpersonal negotiation in a particular environment, and derives its sense of obligation and guilt from the internalization of these processes. Moral reasoning is done by *us*, not by natural selection.

At the same time it should be obvious that human morality cannot be infinitely flexible. Of our own design are neither the tools of morality nor the basic needs and desires that form the substance with which it works. Natural tendencies may not amount to moral imperatives, but they do figure in our decision-making. Thus, while some moral rules reinforce species-typical predispositions and others suppress them, none blithely ignore them.[37]

Evolution has produced the requisites for morality: a tendency to develop social norms and enforce them, the capacities of empathy and sympathy, mutual aid and a sense of fairness, the mechanisms of conflict resolution, and so on. Evolution also has produced the unalterable needs and desires of our species: the need of the young for care, a desire for high status, the need to belong to a group, and so forth. How all of these factors are put together to form a moral framework is poorly understood, and current theories of moral evolution are no doubt only part of the answer.

In the remainder of this book, I will investigate the extent to which aspects of morality are recognizable in other animals, and try to illuminate how we may have moved from societies in which things were as they were to societies with a vision of how things ought to be.

2

SYMPATHY

Any animal whatever, endowed with
well-marked social instincts, the parental and filial affections being here included, would inevitably acquire a
moral sense or conscience, as soon as
its intellectual powers had become as
well developed, or nearly as well developed, as in man.

Charles Darwin[1]

It is simply unimaginable that fish would come to the rescue of an
unlucky pond mate who is jerked out of the water, that they would
bite the angler's line, or butt their heads against his boat in protest.
We also do not expect them to miss their mate, search around, stop
eating, and waste away. Fish are, well, cold to each other. They
neither groom one another like primates, nor do they mutually lick,
nibble, preen, or chat. I say this without any antifish bias. As a
lifelong aquarium enthusiast, I can watch these animals for hours, yet
I would never recommend them to someone in need of affection.

How different from the warm-blooded animals that took to the sea
eighty million years ago!

Warm Blood in Cold Waters

Reports of leviathan care and assistance go back to the ancient
Greeks. Dolphins are said to save companions by biting through
harpoon lines or by hauling them out of nets in which they have
gotten entangled. Whales may interpose themselves between a

hunter's boat and an injured conspecific or capsize the boat. In fact, their tendency to come to the defense of victims is so predictable that whalers take advantage of it. Once a pod of sperm whales is sighted, the gunner need only strike one among them. When other pod members encircle the ship, splashing the water with their flukes, or surround the injured whale in a flowerlike formation known as the marguerite, the gunner has no trouble picking them off one by one. Such "sympathy entrapment" would be effective with few other animals.

But am I justified in using the term "sympathy," which after all is a venerated human concept with very special connotations? Let us for the moment simply speak of *succorant behavior,* defined as helping, caregiving, or providing relief to distressed or endangered individuals other than progeny. Thus, the dog staying protectively close to a crying child shows succorance, whereas the same dog responding to the yelps of her puppies shows nurturance. In reviewing the succorant behavior of animals we will pay special attention to characteristics it might share with human sympathy, the most important being empathy—that is, the ability to be vicariously affected by someone else's feelings and situation. Psychologists and philosophers consider this capacity so central that "empathy" has gradually replaced "sympathy," "compassion," "sorrow," and "pity" in much of their writings (I have even seen the famous Stones' song paraphrased as "Empathy for the Devil").

This blurring is unfortunate, for it ignores the distinction between the ability to recognize someone else's pain and the impulse to do something about it. Administering electrical shocks to someone else's genitals or pouring bleach in open wounds, as done by the torturers of our fine race, involves the very same ability of knowing what makes others suffer, yet it is quite the opposite of sympathy. What sets sympathy apart from cruelty, sadism, or plain indifference is that sensitivity to the other's situation goes together with *concern* about him or her. As neatly summed up by the psychologist Lauren Wispé: "The object of empathy is understanding. The object of sympathy is the other person's well-being."[2]

Whether based on empathy or not, animal succorance is the functional equivalent of human sympathy, expected only in species that know strong attachment. I am not speaking here of anonymous aggregations of fish or butterflies, but the individualized bonding, affection, and fellowship of many mammals and birds.

There certainly is no shortage of attachment among whales and dolphins, which may beach themselves collectively because of their reluctance to abandon a distressed group mate, including a disoriented one. Whereas this action is often fatal to an entire herd, James Porter, an American oceanographer, describes a fascinating exception. When in 1976 thirty pseudorcas (false killer whales) had stranded on an island off the coast of Florida, they remained together in shallow water for three entire days until the largest one had died. The twenty-nine healthy whales would have been unable to return to the ocean (hence would have perished with their apparent leader) under normal tidal conditions. But the tidal range happened to be minimal, so that most of the time the whales did have the option of leaving. "Stranding" is the wrong word, therefore; the whales stayed close to shore of their own accord.

With blood exuding from his right ear, the sick male was flanked and protected by fourteen or fifteen whales in a wedge-shaped configuration. The group was noisy, producing an incredible variety of chirps and squeaks. "With some trepidation, but no common sense," as he commented afterward, Porter entered the water to snorkel toward the group. The outermost individual responded by breaking loose and heading menacingly toward him. Instead of attacking, however, the whale lowered its head and slid underneath the scientist, lifting him out of the water and carrying him to the beach. This procedure was repeated three times, after which Porter tried his luck on the other side. There, too, the outermost whale carried him back to land several times. Porter noted that the whales lost interest in him as soon as he took off his snorkel, suggesting that they were showing a rescue response to sounds that perhaps resembled those of a clogged blowhole.[3]

The U.S. Coast Guard was unable to break up the formation or push the pseudorcas offshore: "If separated from the pack a whale would become highly agitated, and no amount of human effort could restrain its returning to the group. As soon as the whales touched each other, however, they became docile and could easily be nudged into deeper water."[4] (The calming effect of body contact extended to humans who, in a typical act of cross-species sympathy, applied suntan oil to whale backs exposed to sun and air.)

Once the large male had died, the formation around him loosened. Breaking ranks, the whales headed for deeper water while uttering high-pitched descending whistles. Autopsy revealed that the male,

6 meters long, had a massive worm infection in his ear. It is possible that parasitic worms impair a whale's echolocation system, hence its feeding efficiency: the victim's stomach proved empty.

While this account by no means solves the mystery of whale strandings, it gives an idea of how extraordinarily attached to each other these creatures are. If attachment and bonding are at the root of succorant behavior, parental care must be its ultimate evolutionary source. As explained by Irenäus Eibl-Eibesfeldt, with the evolution of parental care in birds and mammals came feeding, warming, cleaning, alleviation of distress, and grooming of the young, which in turn led to the development of infantile appeals to trigger these activities. Once tender exchanges between parent and offspring had evolved— with the one asking for and the other providing care—they could be extended to all sorts of other relationships, including those among unrelated adults. Thus, in many birds the female begs for food from her mate with the same gaping mouth and wing shaking as that of a hungry fledgling, while the male demonstrates his caretaking abilities by providing her with a nice tidbit.

Absorption of parental care into adult human relationships is evident from the widespread use of infantile names (such as "baby") for mates and lovers, and the special high-pitched voice that we reserve for both young children and intimate partners. In this context, Eibl-Eibesfeldt mentions the kiss, which probably derives from mouth-to-mouth feeding of masticated food. Kissing without any transfer of food is an almost universal human expression of love and affection, which, according to the ethologist, resembles kiss-feeding "with one partner playing the accepting part by opening the mouth in a babyish fashion and the other partner performing tongue movements as if to pass food."[5] Significantly, chimpanzees both kiss-feed their young and kiss between adults. A close relative of the chimpanzee, the bonobo, even tongue-kisses.

The continuum of nurturance, attachment, and succorance may not be entirely understood; still, it is hard to argue with its existence. It explains why dolphins and whales, perfectly adapted to the same milieu as fish, act totally unlike them when members of their species are in trouble. A long evolutionary history of parental care combined with a high degree of mutual dependence among adults has endowed these sea mammals with a profoundly different attitude toward one another.

Special Treatment of the Handicapped

For the same reason that whales may fall victim to sympathy entrapment, we humans are vulnerable too. Heini Hediger, the Swiss pioneer of *zoo biology* (the discipline that uses ethology for the improvement of captive conditions), recounts how a lonely gorilla once captured an inexperienced keeper. Having noticed the gorilla desperately struggling to free her arm from the cage bars, the keeper hurried to open her cage and assist in the attempt. The ape, who was not stuck at all, quickly hid behind the door to surprise the keeper. All she did was wrap her arms around him—which, in the case of a gorilla, is guaranteed to restrict a man's movements.

Anecdotes of deception—of which there are many with great apes at the center—tend to be scrutinized for signs of intentionality and planning, yet there is another side to the gorilla's my-arm-got-stuck charade. Hediger comments that the ape must have foreseen the caretaker's humane reaction. The important question arises whether anticipation of helpful behavior to the point of exploitation is possible in a creature devoid of aiding tendencies. In other words, does the setting of a sympathy trap not imply familiarity with the feeling of sympathy?

Perhaps not. Apes may simply learn that one of the many amazing things humans do is take care of distressed individuals, and that this tendency can be used against them. Yet, as we will see, the same tendency exists in other primates, and deception resting on this principle is not limited to human-ape relationships. When in the Arnhem chimpanzee colony the oldest male, Yeroen, had hurt his hand in a fight with an up-and-coming young male, he limped for a week even though the wound seemed superficial. After a while we discovered that Yeroen limped only if he could be seen by his rival. For example, he would walk from a point in front of him to a point behind him, hobbling pitifully at first, only to change to a perfectly normal gait when out of the other male's sight. The possibility that injuries inhibit aggression by rivals may explain Yeroen's attempt to create a false image of pain and suffering.[6]

Primates, including humans, develop succorant tendencies surprisingly early in life, thus contradicting an extensive literature according to which the young are egocentric, mean, even egregiously sinful. This negative view reflects a peculiar assumption about kindness: instead of flowing from the heart—or whatever is taken as the center of emotion—care and sympathy are considered products of the brain.

Because young children lack a high level of cognition and moral understanding, the reasoning goes, there is no way for them to overcome their selfishness. Yet when the psychologist Carolyn Zahn-Waxler visited homes to find out how children respond to family members instructed to feign sadness (sobbing), pain (crying out "ouch"), or distress (coughing and choking), she discovered that children little more than one year of age already comfort others. It is a milestone in their development: an aversive experience in another person draws a concerned response, such as patting, hugging, rubbing the victim's hurt, and so on. Because expressions of sympathy emerge at such an early age in virtually every member of our species, they are as natural an achievement as the first step.

In an attempt to keep the focus on mental processes, Philip Lieberman, an American linguistic anthropologist, interpreted the same result as the first sign of what he chauvinistically labeled "altruism in the higher human sense."[7] Apart from the obvious emotional base, Lieberman emphasized cognition and language, thus ignoring the fact that the actions of one-year-olds far outstrip their verbal abilities. Precisely for this reason Zahn-Waxler has expressed reservations about interviews as a way of gauging empathy and sympathy. With the trouble young children have putting sentiments into words, they may well come across as egocentric when in reality they are already quite caring and protective.

Attention to what children actually do, rather than what they tell interviewers about themselves, is revolutionizing our view of moral development: emotions and actions often seem to come first, rationalizations and justifications later. These insights are also relevant for animal research, which of necessity relies on observation. The same techniques can be applied in both cases, as psychologists unintentionally discovered in their experiments in the household: some pets appeared as upset as the children by the "distress" of a family member, hovering over them or putting their heads in their laps with what looked like great concern. Possibly, then, "altruism in the higher human sense" is not limited to our species.

One day in the rhesus group at the Wisconsin Primate Center, several adult members of the highest-ranking matriline attacked a small juvenile named Fawn, viciously biting their screaming victim from different sides. In typical rhesus fashion, the aggressors regularly turned around to glare at Fawn's mother and sisters, keeping them away from the scene. The fight was so unequal and so severe that I broke it up, shouting at the top of my lungs. (Only a few times a year would we consider such intervention necessary.) Fawn was completely overcome with fear when her attackers finally left. For quite a while she lay on her belly screaming loudly; then she suddenly jumped up and fled. Afterward she sat in a hunched position, looking miserable and exhausted. After two minutes her older sister approached to put an arm around her. When the dazed Fawn failed to respond, her sister gently pushed and pulled at her as if trying to activate her, then embraced her again. The two sisters ended up huddling together while their mother, perhaps in order to avert further problems, groomed the female who had spearheaded the assault.

I must add that comforting acts such as the one by Fawn's sister are rare in a rhesus group. In fact, responses to distressed individuals are most conspicuous in the very young of this species. My assistant, Lesleigh Luttrell, and I would no doubt have missed these instances had we not made a special study of social development. Young infants are easily overlooked; they are tiny, look alike, and seem to matter little in the bigger scheme of things. Our study, however, forced us to recognize individually even the smallest infants, recording their every move. It turned out to be a fascinating experience, because at this early age rhesus monkeys do not yet show the rather intolerant, combative temperament typical of their species.

If one of our little subjects screamed for whatever reason, another infant often hurried over for a brief mount or embrace. An infant might have been threatened or frightened, or might have fallen out of the climbing frame. Most of the time the resulting contact seemed next best to a maternal embrace in terms of calming the infant. Yet whereas a mother's protective and reassuring intentions are rather evident, the same could not be said of the peers. Irresistibly drawn to the distressed infant, other infants showed a strong inclination to make contact, but it was not clear that they had the other's interests in mind.

Once, when an infant had been bitten because it had accidentally landed on a dominant female, it screamed so incessantly that it was soon surrounded by many other infants. I counted eight climbing on top of the poor victim—pushing, pulling, and shoving each other as well as the infant. Obviously the infant's fright was scarcely allevi- ated; the response seemed blind and automatic, as if the other infants were as distraught as the victim and sought to comfort *themselves* rather than the other.

Although it is hard to know the intentions of animals, we can speculate about them and may one day reach the point of testing one interpretation against another. For a research program into animal empathy, it is not enough to review the highlights of succorant behavior; it is equally important to consider the *absence* of such behavior when it might have been expected. Here are two striking examples from rhesus monkeys.

Example 1   Rita broke her arm at the age of seven months, leaving it dangling lifelessly at her side. At first we thought the arm had been disjointed, but x-rays showed it to be fractured. The veterinarian decided not to do anything; bones tend to heal rapidly in young monkeys, and treatment would have required the infant's removal from the group. Remarkably, during the weeks in which Rita could not use her arm, none of the other monkeys seemed to notice her handicap. Juveniles would roughhouse with her as if nothing was the matter, and high-ranking adults would threaten or chase her as they would any other junior nuisance. Even Rita's mother, Ropey, did not in any way alter her behavior; she subjected the infant to exactly the same weaning trauma as other mothers did to Rita's peers.

Rita endured this treatment, and her arm did heal perfectly.

Example 2   A giant vertical wheel in the middle of the cage invites all sorts of games, the most spectacular being pubertal males spinning it faster and faster so as to catapult themselves to the cage ceiling, meters above the wheel. I have never seen a device so effective in keeping monkeys occupied and active. As with all such devices, it is not entirely without danger. One day an infant shrieked in panic when its arm caught in the spokes of the wheel while "teenagers" were playing their daredevil games. An adult male resting next to the wheel became so annoyed by the din that he threatened and hit out at the screaming infant each time it passed. Because this male was normally quite tolerant of small infants, the scene suggested his total lack of realization of the infant's predicament.

In the above examples, monkeys faced an acute situation or a temporary change in another's condition. What if an impairment lasts longer than a couple of weeks? In such cases, monkeys do learn to adjust to the other individual in the same way that they learn to take the needs and inabilities of the young into account. I always admire the complete control of adult males at play; with formidable canines, they gnaw and wrestle with juveniles without hurting them in the least. During play with older and stronger partners, on the other hand, monkeys pull no punches; juveniles will jump on an adult male's back and hit him in the face with an energy that would be fatal if it happened the other way around. Primates play one way with the strong, another way with the weak.[8]

Play inhibitions most likely are a product of conditioning. From an early age, monkeys learn that the fun will not last if they are too rough with a younger playmate; the youngster will scream in protest, try to pull away, or worse, the play will be broken up heavy-handedly by a protective mother. These negative consequences shape the behavior of older individuals. The same process of *learned adjustment* may explain why handicapped members of monkey societies are treated differently. Healthy members do not necessarily know what is wrong, but gradually become familiar with the limitations of their less fortunate mates.

Learned adjustment is best contrasted with *cognitive empathy*. that is, the ability to picture oneself in the position of another individual. This is an extension of sensitivity to expressions of emotion but goes quite a bit further. To explain how it works, imagine that a friend has lost both arms in a car accident. Just from seeing his condition, or hearing about it, we will grasp the reduction in physical ability he has undergone. We can imagine what it is like to have no arms, and our capacity for empathy allows us to extrapolate this knowledge to the other's situation. Our friend's dog, by contrast, will need time to learn that there is no point in bringing her master a stick to fetch, or that the familiar pat on the back is being replaced by a foot rub. Dogs are smart enough to get used to such changes, but their accommodation is based on learning rather than understanding. The result may be somewhat the same. But in the first case, differential treatment of the disabled is based on an understanding of their limitations; in the second, on familiarity with their behavior. Needless to say, the dog's learned adjustment is a slower process than cognitive empathy.[9]

Cognitive empathy may not be widespread in the animal kingdom. It occurs in people and perhaps in our closest relatives, the apes, but may be absent in other animals. There are as yet no indications that the special treatment that monkeys accord the handicapped is based on anything other than learned adjustment. The attitude toward Mozu is a case in point: the tolerance toward her probably reflects learning by other monkeys in Jigokudani Park that Mozu is slow and poses no threat. Nothing in their behavior suggests that they realize how or why Mozu differs from them, and what her special needs are.

Below are three further examples of adjustment by monkeys to permanently disabled species members.

Azalea: an infant with a chromosomal abnormality In 1988, a unique rhesus monkey was born at the Wisconsin Primate Center. Azalea had a strange, somewhat vacuous facial expression and was clearly deficient in motor abilities. Her condition, which had come about spontaneously, was detected early on because she happened to be part of our developmental study. Testing showed that she had the extremely rare condition of autosomal trisomy: instead of possessing only pairs of chromosomes, she had one triplet, similar to the chromosomal condition underlying Down syndrome in humans. In a further parallel, Azalea was born to an aging, almost postreproductive female (more than twenty years old) who previously had produced eleven normal offspring.

Running, jumping, and climbing posed major challenges that Azalea achieved at a considerably later age than her peers, and then only incompletely. Her coordination was imperfect, her reaction time slow. She was the only monkey who never learned to walk through the spinning wheel by stepping into it at one point and exiting at another point as she took changes in speed and position into account (the way we learn to step smoothly in and out of a revolving door). If Azalea was "trapped" in the wheel, an elder sister often rescued her by pulling her out. Feeding was another serious problem: instead of holding hard monkey biscuits in her hands and biting off pieces, Azalea licked crumbs from the floor in doglike fashion until the age of five months. She would never have survived in the wild.

Socially, Azalea was aberrant as well. She moved out of her mother's reach for the first time on the thirty-seventh day of life, compared with an average age of thirteen days for her peers. Azalea's mother never showed signs of rejection, yet was not particularly interested in this offspring. Azalea's elder juvenile sister, on the other hand, did pay extra attention, carrying her around well beyond the normal age for such sisterly care and protecting her against other monkeys. If others acted toward Azalea in a manner to which any normal rhesus infant would object, such as pulling out one hair after another during grooming, her sister would interrupt the activity even though Azalea herself uttered not the slightest protest.

Although Azalea's membership in the top matriline offered strong protection, she was at the receiving end in a few serious fights—because of miscalculation. If her mother threatened another monkey, Azalea would join in almost blindly, enthusiastically grunting and lunging at the same individual, however big he or she happened to be. Alliance formation (two parties banding together against another) is common in monkey groups; it often results in large-scale confrontations in which many parties go after one another. On these shifting battlegrounds it is of paramount importance for each monkey to keep track of the positions and movements of all others, friend as well as foe. Inasmuch as this task was totally beyond Azalea's capacities, involvement in her mother's battles often placed her in an uncovered position face to face with an adversary who could seriously harm her. Her opponents rarely hesitated, and one time they literally wiped the floor with her. An adult male dragged her down the stone incline of the enclosure, smashing her head onto the concrete floor. Azalea went into a seizure of several seconds, during which she lost control on one side of her body. She did learn from this incident, though: afterward she kept a wary eye on this particular male, avoiding him even if he showed up only in the distance.

Most of the time, Azalea was exceptionally passive. She rarely played with others; if she did, it was with younger partners, her peers being too fast on their feet and too wild. It is a testimony to the group's complete acceptance of this backward juvenile that others, including nonrelatives, groomed Azalea twice as much as they did her peers.[10]

At the age of thirty-two months, Azalea showed severe convulsions and disorientation (bumping her head into walls) and had to be put down.

Wania-6672: an infant with a neural disorder   In 1972 two Canadian primatologists, Linda and Laurence Fedigan, observed an unusual infant born into a population of Japanese macaques in a 44-hectare brushland enclosure in Texas. The infant showed symptoms of cerebral palsy. Having more control over arms than legs, Wania-6672 would hop like a rabbit, swinging both legs simultaneously forward outside his arms. The infant appeared visually impaired: he always sniffed at other monkeys as if to identify them, and regularly collided with bushes and cactus. Another similarity with palsied human children was excessive aggressivity and hyperactivity. Wania-6672 showed extreme hostility toward monkeys grooming his mother, trying at all costs to insert himself between the two parties.

He spent an enormous amount of time following his mother, seeking to nurse, and was often carried and groomed by her. When they were separated Wania-6672 vocalized persistently; he never achieved any degree of independence. Juveniles, particularly females, frequently ran over to cradle and hold him when he gave distress calls.

Group members were thoroughly puzzled by this infant's behavior. Instead of running away when threatened, Wania-6672 would scream and crash aimlessly about. "When confronted with this strange response, most monkeys stopped threatening immediately and many peered at the abnormal infant as if confused by the unusual response."[11] Perhaps because of the lack of appropriate behavior, other monkeys lost interest in trying to teach him the usual rules of conduct; they tended to ignore misbehavior by Wania-6672 that they would never accept from other infants. One time an adult male did nothing when Wania-6672 collided head on with the adult's groin. Another time, the alpha male was being groomed by a female, reclining and dreaming away until the infant began tripping over his feet. The male raised his eyebrows several times—a sign of irritation—then sat up staring at the offender with a full-blown threat face. But he lay down immediately when he saw that it was "only" Wania-6672. Any other infant would have been grabbed by the scruff of its neck and rubbed in the dirt.

I585-B: a congenitally blind infant   In 1982 a blind infant was born into a free-ranging population of rhesus monkeys released onto a Caribbean island. Apart from being sightless, the infant appeared perfectly normal: he played, for instance, as much as other infants his age. Compared to his peers, I585-B often broke contact with his mother, thereby placing himself in situations that he could not recognize as dangerous. His mother responded by retrieving and restricting him more than other mothers did with their infants.

In other studies of blind infant monkeys such infants were never left alone, and specific group members stayed with them whenever the group moved.[12] Similarly, I585-B's kinship unit was extremely vigilant and protective, as noted by Catherine Scanlon.

On several occasions, the blind infant was seen in the lower branches of a tree between two and five meters from close kin: the approach of an unrelated animal was observed to result in an unusually high number of threats from one or more of these kin, in particular his mother, his aunts, and his five-year-old male cousin.[13]

Let us not forget that, in theory, learned adjustment can take *two* directions. The first is exploitation, as when healthy monkeys learn to take advantage of the lack of speed, strength, and perception of a disabled group mate. For example, the younger members of the stump-tail monkey group at the Wisconsin Primate Center used to tease Wolf, an old, virtually blind female, by jumping or climbing on her back—a feat they would not have dared before Wolf lost her eyesight.

Among animals who have lived together for a long time, however, such maltreatment is not the rule. What is most striking about adjustment to handicapped monkeys is that it so often takes the opposite form: instead of being torn to pieces or abandoned as a useless community member, the disabled individual receives extra tolerance, vigilance, and care. This pattern applied to Wolf as well: after the deterioration of her eyesight, the group's adult males became extremely protective. Each time human caretakers tried to move the monkeys from the indoor to the outdoor section of the enclosure, adult males would stand guard at the door between the sections, sometimes holding it open, until Wolf had gone through. Interposing themselves between her and the caretakers, they made it quite obvious that Wolf still meant a great deal to them.

Special treatment of the handicapped is probably best regarded as a combination of learned adjustment and strong attachment; it is the attachment that steers the adjustment in a positive, caring direction. Sometimes, however, special treatment occurs when there has been little or no time for learning. For example, monkeys may suddenly increase their vigilance if one among them is injured or incapacitated. When a juvenile in a captive baboon colony had an epileptic seizure, other baboons immediately turned highly protective. An elder brother placed his hand on the disabled juvenile's chest, threatening people who wanted to enter the compound for a closer look. According to Randall Keyes, who reported the incident, this brother normally did not act so protectively.[14]

Such an immediate response to a group mate's vulnerability is not easily explained on the basis of a slow learning process. Perhaps primates follow a hard and fast rule—without too much reflection—that tells them to double their protectiveness as soon as one among them fails to respond to danger. Or they may have learned that individuals who remain immobile at critical moments tend to get into trouble, and they generalize this knowledge to a newly incapacitated group mate. Explanations along these lines may account for the extra protection of the blind infant, Wolf, and the baboon with a seizure.

The alternative would be that they recall their *own* encounters with approaching danger when watching another's lack of response. This possibility would be of the utmost interest, because to understand another's predicament based on one's own experiences would require extrapolation from self to other.

Responses to Injury and Death

If attachment underlies care and sympathy, the attitude toward dead or dying companions is worth investigating because there is no more poignant evidence for attachment than agony following the final breath of a relative or companion. A well-known example is elephants, who sometimes pick up the ivory or bones of a dead herd member, hold the pieces in their trunks, and pass them around. Some pachyderms return for years to the spot where a relative died, touching and inspecting the relics. Do they miss the other? Do they recall how he or she was during life?

Dereck and Beverly Joubert watched the final hours of an old bull lying in the sands of the Kalahari Desert. Other elephants tried to raise the dying individual by slipping their trunk and tusks underneath him and lifting as hard as they could, some breaking ivory in the effort.

Cynthia Moss, in *Elephant Memories,* describes the response of elephants in Amboseli National Park when a poacher's bullet entered the lungs of a young female, Tina. After the herd had escaped from danger, Tina's knees started to buckle, and the others leaned into her so as to keep her upright. She slipped beneath them nonetheless, and died with a shudder.

Teresia and Trista became frantic and knelt down and tried to lift her up. They worked their tusks under her back and under her head. At one point they succeeded in lifting her into a sitting position but her body flopped back down. Her family tried everything to rouse her, kicking and tusking her, and Tallulah even went off and collected a trunkful of grass and tried to stuff it into her mouth.[15]

Afterward, the others sprinkled earth over the carcass, then went off into the surrounding bushes to break off branches, which they placed over Tina's body. By nightfall the corpse was almost completely covered. When the herd moved off next morning, Teresia was the last one to leave. Facing the others with her back to her dead daughter, she reached behind herself and felt the carcass with her hind foot several times before she very reluctantly moved off.

Monkeys react to the death or disappearance of an attachment figure in a way outwardly similar to human grieving. In the 1960s this process was investigated in the laboratory by Charles Kaufman and Leonard Rosenblum, who separated infant monkeys from their mothers. A *protest phase,* with vigorous calling and searching for the mother, was followed by a *despair phase,* with nonresponsiveness, loss of appetite, slouching posture, and empty gaze. It is unclear, however, if such depressions ever reach the point where monkeys *die* of grief in the way that Jane Goodall described for the wild chimpanzee Flint, who died three weeks after the loss of his mother, the famous Flo. Flint was already eight and a half years old, but had been unusually dependent. In Goodall's words, "His whole world had revolved around Flo, and with her gone life was hollow and meaningless."[16] Autopsy revealed inflammation of Flint's stomach and abdomen. It is indeed possible that his death was caused by a weakened immune system due to depression, yet we cannot rule out the obvious alternative that Flo and Flint had fallen victim to the same disease, and that Flint had merely held out a little longer than his mother.[17]

If the reverse happens, that is, if a mother loses her infant, primates demonstrate their attachment by carrying the corpse around for days until it literally falls apart. Once they have given up the carcass, it usually is hard to recognize signs in female monkeys, other than swollen breasts, of the loss of an infant. Yet some females are restless and seem to look for their offspring. I have known rhesus females who paced and vocalized for an entire day following removal of the cadaver. And Barbara Smuts, an American primatologist, describes in *Sex and Friendship in Baboons* how a wild female baboon named Zandra responded to the loss of her three-month-old infant, Zephyr, who had been severely bitten, most likely by a male baboon.

For the next several days, I looked for signs of grief in Zandra but could see none. During this time, the troop did not travel near the spot where the infant had been killed and where she had lost the body. Then one day, about a week after her infant died, the troop passed through the same bushy area. As they drew near, Zandra became extremely agitated. She rushed about as if looking for Zephyr, and then she climbed a tree. When she came to the top, she looked all around and began calling . . . Her searching activities were unmistakable, and she repeated her agitated looking and calling each time the baboons passed through this area for the next few weeks.[18]

Similarly, female chimpanzees commonly wail and whimper, and sometimes burst out screaming after the loss of an offspring. One of the females in the Arnhem colony, confusingly named Gorilla because of a resemblance to the other ape species, would not only whimper and scream, but after each flareup would vigorously rub both eyes with her knuckles exactly like a human child drying tears. Because humans are supposedly the only primates to shed tears when they cry, Gorilla's behavior so surprised me that I requested that her eyes be checked for infection. The veterinarian could not find anything, however, and I must stress that I never actually saw tears.[19]

Seeing the termination of a familiar individual's life, chimpanzees may respond emotionally as if realizing, however vaguely, what death means—or at least that something terrible has befallen the other. In a second incident involving Gorilla a young adult female, Oortje, simply dropped dead. Oortje had been apathetic since delivering, two months earlier, a healthy son whom she rejected. She had started coughing before that time, and her condition had deteriorated despite treatment. Kept indoors because of the winter, the colony had been divided into two groups that could hear but not see each other. At a certain moment one day Jeanne Scheurer, a regular zoo visitor who knew every chimpanzee by name, saw Gorilla intently stare at Oortje, who was sitting on a log. Without apparent reason, Gorilla burst out screaming in an excited, hysterical voice. The screaming was neither aggressive in tone nor accompanied by threat gestures; rather, Gorilla seemed disturbed by something she saw in Oortje's eyes or demeanor. Oortje herself, silent until this point, now feebly screamed back at Gorilla, then tried to lie down, fell off the log, and remained motionless on the floor. All this happened within a minute. Heart massage and an attempt at mouth-to-mouth resuscitation by the caretaker (after having moved the other chimpanzees inside) were in vain. A female in the other hall uttered screams similar in sound to Gorilla's, then every chimpanzee in the building went completely silent. Oortje must have died of heart failure; autopsy showed a massive infection of heart and abdomen.

As mentioned before, one adult male in Arnhem did not survive serious injuries contracted in a fight. After hours of surgery we placed him in one of the night cages, where he died. That evening, when the rest of the colony entered the building, there was absolute silence. The next morning, silence persisted even when the keeper arrived with food (normally a time for tumultuous joy). Vocal activity resumed only after the corpse had been removed from the building.

One early morning in 1968, Geza Teleki followed a small party of wild chimpanzees shortly after they had descended from their sleeping nests. He hurried over to a site from which raucous calling could be heard. Six adult males were wildly charging about, their wraaah calls echoing off the valley walls. Activity was concentrated around a small gully in which the motionless body of a male, Rix, was sprawled among the stones. Although Teleki had not seen the body drop—at the critical moment he had stepped into a safari ant nest—he felt he was witnessing the very first reaction to Rix's neck-breaking fall out of a tree. Several individuals paused to stare at the corpse, after which they redirected their tension by vigorous charging displays away from it, hurling big rocks in all directions. Amid the din chimpanzees were embracing, mounting, touching, and patting one another with big, nervous grins on their faces.

Later on, some chimpanzees spent considerable time staring at the body. One male leaned down from a limb, watched the corpse, then whimpered. Others touched or sniffed Rix's remains. An adolescent female uninterruptedly gazed at the body for *more than an hour,* during which she sat motionless and in complete silence. After three hours of activity around the corpse, one of the older males finally left the clearing, walking downstream along the valley bottom. Others followed one by one, glancing over their shoulder toward Rix as they departed. One male approached the remains, leaned over for a final inspection, then hurried after the others.

When a companion is close to death, chimpanzees act as if they are aware of the pain of this individual, as movingly described six decades ago by Robert Yerkes, who watched a group of juveniles.

Impressive indeed is the thoughtfulness of the ordinarily care-free and irresponsible little chimpanzee for ill or injured companions. In the same cage were a little male and two females, one of the latter mortally ill. She was so ill that much of the time she lay on the floor of the cage in the sunlight, listless and pathetic. There was excellent opportunity to observe the attitude of her lively companions towards this helpless invalid. In all their boisterous play they scrupulously avoided disturbing her, and, in fact, seldom touched her as they climbed, jumped, and ran about the cage. Now and then one or the other would go to her and touch her gently or caress her; or again one of them, fatigued or worsted in some game, would obviously seek refuge and respite by going close to her. In this position, safety from disturbance was assured.

To this, Yerkes adds his interpretation: "A certain solicitude, sympathy, and pity, as well as almost human expression of consideration were thus manifested by these little creatures."[20]

Note that this reading goes well beyond the learned adjustment hypothesis discussed earlier: Yerkes assumes empathy. The same assumption shines through the following account of extraordinary care by Lucy, a chimpanzee in a human family. Maurice Temerlin, an American psychotherapist, analyzed Lucy's every move, personality trait, and sex game from the time she was a baby until she had grown into a 40-kilogram humanized chimpanzee. Lucy was particularly tender with Temerlin's wife, Jane, as described in *Lucy: Growing up Human*.

If Jane is distressed, Lucy notices it immediately, and attempts to comfort her by putting her arm about her, grooming her, or kissing her. If I am the cause of the distress, for example, if we are arguing, Lucy will attempt to pull us apart or to distract me so that Jane's distress is alleviated. If Jane is sick, Lucy notices it immediately. For example, on every occasion when Jane was ill and vomited, Lucy became very disturbed, running into the bathroom, standing by Jane, comforting her by kissing her and putting her arm around her as she vomited. When Jane was sick in bed Lucy would exhibit tender protectiveness toward her, bringing her food, sharing her own food, or sitting on the edge of the bed attempting to comfort by stroking and grooming her.[21]

It is entirely possible that succorant behavior and sensitivity to the needs of others are better developed in apes than in monkeys. From my experience, the death of an individual to whom they are not closely attached produces a rather unremarkable response in monkeys, and I have never heard of monkeys offering the sort of solicitous attention to indisposed companions described above for chimpanzees. While there may be a real difference between monkeys and apes, our knowledge leaves much to be desired. Monkeys do show many kinds of gentle care, protectiveness, and vigilance toward the sick and injured, and most of the time the similarities with apes are more striking than the differences.

The assumption of higher mental processes in chimpanzees today is so prevalent that widespread mammalian habits, such as the cleaning of wounds, are sometimes interpreted differently depending on whether they are exhibited by a chimpanzee or another animal. Wolfgang Dittus documented how toque macaques in Sri Lanka lick each other's wounds caused by fights, a behavior that is especially important for wounds out of the victim's own reach. Because of the antiviral and antibacterial properties of saliva, cuts as long as 15 centimeters heal quickly and leave virtually no scar. The care is so crucial that one male injured upon emigration to another monkey group, without friends to provide this service, returned to his former group—where his wounds were tended by male peers with whom he had grown up.

Dittus does not speculate about his monkeys' awareness of each other's pain, but Christophe Boesch, describing identical behavior in chimpanzees in Taï National Park, Ivory Coast, takes wound cleaning to mean that his apes "are aware of the needs of the wounded" and demonstrate "empathy for the pain resulting from such wounds."[22] Other chimpanzees lick away blood, carefully remove dirt, and prevent flies from coming near the wounds. They also protect injured individuals, and travel more slowly if these individuals cannot keep up the pace.

Does this indeed prove that chimpanzees have empathy, and, by extension, that monkeys and other injury-cleaning mammals do too? Unfortunately, the tending of wounds per se tells us nothing about the underlying mental processes. A skeptic could argue that it only proves that blood tastes good; indeed, it is not unusual for primates to lick blood off plants or branches splattered during a fight or a birth. I believe there is more to the cleaning of wounds; many hard-to-convey details in the behavior of chimpanzees (the way they approach an injured individual; the concerned look in their eyes; the care they take not to hurt) make me intuitively agree with the views of Yerkes, Temerlin, and Boesch. That such details are less obvious in monkeys does not necessarily mean that they perform the same actions devoid of feelings, or without any understanding of what happened to the victim; monkeys are harder to read because of their greater evolutionary distance from us.

That high levels of care are not restricted to species close to us has already been demonstrated with salient examples from elephants and dolphins; the same applies *within* the primate order. Some ethologists, myself included, would no doubt have thought caring responses to be less developed in prosimians, an ancient branch of the order from which the true simians evolved. In a few languages prosimians are known as *half-monkeys;* they have a reputation for being primitive and unsophisticated compared to monkeys. This image may need correction in light of what Michael Pereira, an American expert on their behavior, described to me as an "absolutely mind-blowing" series of events among ring-tailed lemurs in a forested enclosure at the Duke University Primate Center.

One day a three-month-old infant climbs up an electric fence, receives a jolt to her temple, and falls to the ground in convulsions. The student assistant, Louis Santini, who witnesses the drama, runs off to seek help, and returns to find the infant on the back of her grandmother, who normally *never* carries her. The infant's own mother, who had not been present at the incident, pays no attention and continues to feed in a tree in the distance. The grandmother carries the infant for ten minutes, then drops her off at a quiet spot where the infant sits in a daze.

After another ride on the grandmother's back and continued lack of interest by the mother, the infant is approached by peers, who groom her intensely for several minutes—another highly unusual behavior, as infants rarely groom anyone. Three infants start a round-robin game among themselves near the dazed one, occasionally interrupting their play to take turns briefly grooming the victim. When the group of lemurs moves off, the infant climbs on the back of her mother, who has not been near her for over an hour, and succeeds in riding on her. The mother, known as rather rejective, suddenly sits up to violently shake the infant off her back. The grandmother responds instantly by attacking the mother, which results in the mother's allowing her daughter to remount and stay on for a longer distance. Five minutes later, the group settles down. The infant rests on her mother's ventrum like other infants in the group. She now appears fully recovered (and indeed survives without noticeable damage).

This incident shows that even lemurs immediately recognize when one of them is in trouble and respond appropriately. The victim's peers seemed to notice something wrong with their buddy, and certainly the grandmother went out of her way to offer unusual care, going so far as to discipline her own daughter for not accommodating the injured family member. Could the difference in response between mother and grandmother have had something to do with the fact that only the grandmother had witnessed the mishap? The grandmother's action against her daughter was all the more remarkable because older lemur females almost never meddle in conflicts between daughters and grandchildren. What intrigues me most is that she seemed to teach her daughter how she *ought* to behave; precisely the kind of social pressure viewed in moral terms if seen in humans. Mind-blowing, indeed!

If science has not yet gauged the full depth of the emotional and cognitive life of monkeys and prosimians as expressed in care for distressed individuals, even more remains to be discovered with regard to the apes. Take an observation made repeatedly during reconciliations among both bonobos and chimpanzees, yet never in any monkey species that I have studied with equal intensity: after one individual has attacked and bitten another, he or she may return to inspect the injury inflicted. The aggressor does not accidentally detect the wound, but knows exactly where to look for it. If the bite was aimed at the left foot, the aggressor will without hesitation reach for this foot—not the right foot or an arm—lift and inspect it, then begin cleaning the injury. Such action suggests an understanding of cause and effect and may require the ape to take the other's perspective, as if realizing the impact of his own behavior on someone else.[23]

Chimpanzees also excel at so-called *consolation*. Whereas monkeys do on rare occasions reassure a victim of aggression (witness the incident with Fawn and her sister), such behavior is exceptional. A study of three macaque species by the Italian ethologist Filippo Aureli and his coworkers failed to demonstrate regular care for recent victims of aggression. If anything, others tended to stay away from them, probably because of the risk of further attacks on these same individuals. Quite the contrary happens in the chimpanzee colony at the Yerkes Field Station. Watching the colony from high in a tower with an excellent overview of the compound, research technician Michael Seres enters every observed event directly into a computer. These records show that once the dust has settled after a fight, combatants are often approached by uninvolved bystanders. Typically the bystanders hug and touch them, pat them on the back, or groom them for a while. These contacts are aimed at precisely those individuals expected to be most upset by the preceding event.[24]

The chimpanzee's responsiveness to distress in others goes together with a range of humanlike expressions, unmatched by any other nonhuman primate, to *seek* contact and reassurance. When upset, chimpanzees pout, whimper, yelp, beg with outstretched hand, or impatiently shake both hands so that the other will hurry and provide the calming contact so urgently needed. If all else fails, chimpanzees resort to their ultimate weapon, the *temper tantrum*. They lose control, roll around screaming pathetically, hitting their own head or beating the ground with their fists, regularly checking the effects on the other. Tantrums are commonly thrown by juveniles of three or four years when their mothers' willingness to nurse is on the decline; but fully grown chimpanzees may do so too, for example, when another chimpanzee has refused to share food, or after a confrontation with a major rival has been lost.

I still recall the spectacular tantrums of the old leader of the Arnhem colony, Yeroen, when he was being dethroned by younger and stronger males. He would tumble down the trunk of a tree, as if he could barely hold on, screaming mightily, and begging with outstretched hands to anyone in sight, particularly adult females who could help him chase off his opponent. Walking through the zoo, I could hear Yeroen's pleas for help nearly a kilometer away! Often females or juveniles would go over to Yeroen to place an arm around him and calm him down.

As a final proof of the chimpanzee's sensitivity to the distress of others, compare the total lack of adjustment by Ropey, a rhesus monkey, to her daughter's broken arm, with the following response of a chimpanzee mother. During tensions among adult males in the Arnhem colony, one male picked up a juvenile, Wouter, swinging him furiously above his head and against a wall. The action was not necessarily directed against Wouter himself; chimpanzee males enhance their charging displays with virtually anything that makes noise, and juveniles quickly learn to stay out of their way. Several adult females came to Wouter's rescue, and succeeded in putting an end to the molestation, but Wouter limped for weeks afterward.

Fortunately for him, Wouter's mother did seem to understand that something was wrong. Having a younger brother who was still being nursed, Wouter had gotten used to being second in line for maternal attention. This lesson can be hard to learn, requiring mothers to insist by force. During the time that Wouter was disabled, his mother relaxed the rules: on many occasions she shoved her younger son aside, ignoring his fuss, to make room for and tenderly hold the older sibling.

Having Broad Nails

For decades students of animal behavior considered it wrong and naive to speak of animals as wanting, intending, feeling, thinking, or expecting beings. Animals just behave; that is all we know, and all we will *ever* know about them.

Curiously, the key to behavior was sought not within but outside the individual. The individual was merely a passive instrument of the environment. Psychologists studied how responses to stimuli increase if rewarded, and biologists analyzed how behavior spreads if it promotes reproduction. The first is a learning process, the second natural selection; the timescales are of course vastly different, yet the role of the environment as final arbiter of the suitability of behavior was the same. With biologists and psychologists quibbling endlessly over whose discipline offered the better explanation, it is hard to believe that they shared so much common ground.

The critical insights of both disciplines steered attention away from the acting agents themselves. If the environment controls behavior, why do we need the individual? Psychologists came up with their infamous black box, which mediates between stimulus and response, yet remains inaccessible to science. Biologists described animals as survival machines and preprogrammed robots, another way of saying that we should not worry too much about what goes on in their heads. The final presentation of the late B. F. Skinner, who firmly kept the lid on the black box, therefore could hardly have been accidental. Addressing fellow psychologists, he compared cognitive psychologists to creationists, thus throwing together the enemies of behaviorism and Darwinism![25]

When evaluating the succorant behavior of animals, we face already so many obstacles relative to the scarcity of data and the design of experiments that we do not need the additional burden of a narrow-minded rejection of the entire problem of cognition. Some biologists will point out that most of the accounts of caregiving discussed earlier—certainly the more striking ones—concern kin. Why not simply view these instances as investments of genetic relatives in one another, and leave them at that? While perfectly valid when it comes to evolutionary explanations, this point has absolutely no bearing on the question at hand. We are concerned here with motivations and intentions. Regardless of how care is being allotted, the caregiver must be sensitive to the situation of the other, feel an urge to assist, and determine which actions are most appropriate under the circumstances.

If Tallulah stuffs grass into the mouth of a dying herd member, if a chimpanzee goes to hug another who has just been beaten up, or if the top male of a monkey group fails to punish a brain-damaged infant for bothering him, we want to know what makes these animals react in this way. How do they perceive the distress or special circumstances of the other? Do they have any idea of how their behavior will affect the other? These questions remain exactly the same whether or not the other is a relative.

Times are changing. Interest in the mental life of animals is regaining respectability. Whereas some scientists propose a gradual shift in this direction, depending on evidence along the way, others are less patient. Believing it unfair to hold a new perspective hostage to the availability of final answers, they advocate a clean break with the Cartesian view of animals as automatons.

This is not to say that all we need to do is "feel as one" with animals without some critical distance, without putting ideas to the test, and without choosing our words carefully. Discussions about animal behavior often boil down to discussions about language. The ethologist inevitably borrows concepts from common language, which is primarily designed for communication about people. Yet the familiarity of these concepts by no means absolves us from the obligation to be specific about what they mean when applied to animals. Anthropomorphism can never replace science.

Take "reconciliation" and "consolation," two blatantly anthropomorphic terms applied to primate behavior. They refer to circumscribed encounters and come with a set of predictions that, if contradicted, should spell the end of their use. For example, reconciliation is defined as a reunion between former opponents shortly after an aggressive conflict between them. If it were found that reconciliations thus defined do not occur, or do nothing to reduce renewed hostility, it would be time to rethink the label. The same argument applies to older, more widely accepted terms, such as "threat," "greeting," "courtship," and "dominance," which have already gone through a process of fine-tuning and critical evaluation, but can still be questioned at any moment.[26]

It is this use of anthropomorphism as a *means* to get at the truth, rather than as an end in itself, that sets its use in science apart from use by the layperson. The ultimate goal of the scientist is emphatically *not* to arrive at the most satisfactory projection of human feelings onto the animal, but rather at testable ideas and replicable observations. Thus, anthropomorphism serves the same exploratory function as that of intuition in all science, from mathematics to medicine. As advocated by Gordon Burghardt:

> What I am calling for is a critical anthropomorphism, and predictive inference that encourages the use of data from many sources (prior experiments, anecdotes, publications, one's thoughts and feelings, neuroscience, imagining being the animal, naturalistic observations . . . et cetera). But however eclectic in origin, the product must be an inference that can be tested or, failing that, can lead to predictions supportable by public data.[27]

But what about the cherished principle of parsimony—the one great bulwark against all this liberal thinking? The problem is that insofar as monkeys and apes are concerned, a profound conflict exists between *two* kinds of parsimony. The first is the traditional canon that tells us not to invoke higher capacities if the phenomenon can be explained with lower ones. This favors simple explanations, such as learned adjustment, over more complex ones, such as cognitive empathy.

The second form of parsimony considers the shared evolutionary background of humans and other primates. It posits that if closely related species act the same, the underlying process probably is the same too. The alternative would be to assume the evolution of divergent processes for similar behavior; a highly uneconomic assumption for organisms with only a few million years of separate evolution. If we normally do not propose different causes for the same behavior in, say, tigers and lions, there is no good reason to do so for humans and chimpanzees, which are genetically as close or closer.

In short, the principle of parsimony has two faces. At the same time that we are supposed to favor low-level over high-level cognitive explanations, we also should not create a double standard according to which shared human and ape behavior is explained differently.

Such "evolutionary parsimony" is a factor especially when both humans and apes exhibit traits not seen in monkeys, and two explanations are proposed where one may do. If accounts of human behavior commonly invoke complex cognitive abilities—and they certainly do—we must carefully consider whether these abilities are perhaps also present in apes. We do not need to jump to conclusions, but the possibility should at least be allowed on the table.

Behind the debate about parsimony towers the much larger issue of humanity's place in nature. To this day, those who see our species as part of the animal kingdom continue to lock horns with those who see us as separate. Even authors with a distinctly evolutionary perspective often cannot resist searching for the one BIG difference, the one trait that sets us apart—whether it is opposable thumbs, toolmaking, cooperative hunting, humor, pure altruism, sexual orgasm, the incest taboo, language, or the anatomy of the larynx. Countless book titles reflect this search: *Man the Tool-Maker, Man the Hunter, The Ethical Animal, Uniquely Human,* and so on.

Claims of human uniqueness go back to the debate between Plato and Diogenes about the most succinct definition of the human species. Plato proposed that humans were the only creatures at once naked and walking on two legs. This definition proved flawed, however, when Diogenes brought a plucked fowl to the lecture room, setting it loose with the words, "Here is Plato's man." From then on the definition included "having broad nails."[28]

In 1784 Johann Wolfgang von Goethe triumphantly announced that he had discovered the cornerstone of humanity: a tiny piece of bone in the human upper jaw known as the *os intermaxillare.* The bone, though present in other mammals including apes, had long been thought absent in us and had therefore been labeled a "primitive" trait by a Dutch anatomist, Petrus Camper. Goethe's bone, as it became known, confirmed our continuity with nature long before Darwin formulated his theory of evolution. It was a slap in the face—the first of many—to people postulating human uniqueness.

Do claims of uniqueness in any way advance science? Are they even scientifically motivated? Until now, all of these claims have either been forgotten, like Camper's, or required qualification, like Plato's. As a separate species, humans do possess distinct traits, yet the overwhelming majority of our anatomical, physiological, and psychological characteristics are part of an ancient heritage. Holding the magnifying glass over a few beauty spots (our distinct traits are invariably judged advanced and superior) is a much less exciting enterprise, it seems to me, than trying to get a good look at the human animal as a whole.

In this broader perspective, peculiarly human traits are juxtaposed with the obvious continuity with the rest of nature. Included are both our most noble traits and the ones of which we are less proud, such as our genocidal and destructive tendencies. Even if we like to blame the latter on our progenitors (as soon as people hack each other to pieces they are said to be "acting like animals") and claim the former for ourselves, it is safe to assume that both run in our extended family.

There is no need to launch probes into space in order to compare ourselves with other intelligent life: there is plenty of intelligent life down here. It is infinitely more suitable to elucidate the working of our minds than those of whatever extraterrestrial forms might exist. In order to explore earthly intelligences, we need breathing room in the study of animal cognition: freedom from traditional constraints that tell us that nothing is there, or that *if* anything is there, we will never be able to catch even a glimpse.

Critics say there is no way to see what goes on inside an animal's head. That is, of course, not literally what cognitive ethologists are trying to do. Rather, they seek to *reconstruct* mental processes in much the way the nuclear physicist "looks inside" the atom by testing predictions based on a model of its structure. Admittedly, the use of anthropomorphism and anecdotal evidence, along with reservations about the principle of parsimony, have created uncertainty and confusion—as well as lively debate.[29] Yet these are only the delivery pains of a much-needed change in the study of animal behavior.

With this in mind, let us return to succorant behavior and the evolution of empathy and sympathy.

The Social Mirror

Without access to slate or metallic surfaces, and lacking rivers that provide clear reflections, the Biami, a Papuan tribe in New Guinea, were assumed never to have seen their own images. This made them ideal subjects for Edmund Carpenter, a visual anthropologist, out to document people's very first reactions to mirrors.

> They were paralyzed: after their first startled response—covering their mouths and ducking their heads—they stood transfixed, staring at their images, only their stomach muscles betraying

great tension. Like Narcissus, they were left numb, totally fasci-nated by their own reflections; indeed, the myth of Narcissus may refer to just this phenomenon. In a matter of days, however, they groomed themselves openly before mirrors.[30]

Polaroid pictures proved even more puzzling. At first, the Biami failed to understand: the anthropologist had to teach them how to read the picture by pointing to the nose, then touching the real nose, and so on with other parts of the body. With recognition came fear. The individual in the picture would tremble uncontrollably, turn away from the photograph, then slip away to a private spot with the picture firmly pressed against his chest. There he would stare at it, without moving, for up to twenty minutes. Because of this response, Carpenter speaks of the "terror of self-awareness." But this stage was quickly left behind. Within days villagers were happily making movies of themselves, taking snapshots of each other, listening to themselves on a tape recorder, and proudly wearing a self-portrait on their forehead.

Of course, the Biami had not been without self-consciousness be-fore the anthropologist set foot in their village. The only thing that mirrors and pictures do is bring awareness of oneself into sharper focus and demonstrate its presence to the outside world. Had these people not responded with surprise, fear, and fascination, we might have thought either that they had seen themselves before, or that they failed to grasp what they saw. The latter is unimaginable, as self-awareness is at the core of being human. Without it we might as well be folkloric creatures without souls, such as vampires, who cast no reflections. Most important, we would be incapable of cognitive em-pathy, as this requires a distinction between self and other and the realization that others have selves like us.

No wonder responses to mirrors have attracted the attention of students of animals as well. While almost all visually oriented mam-mals initially try to reach or look behind a mirror, only two non-human species—chimpanzees and orangutans—seem to understand that they are seeing themselves. The special status of these apes has been recognized for a long time. In 1922 Anton Portielje, a Dutch naturalist, remarked that, whereas monkeys fail to understand the relation between their reflections and themselves, an orangutan "at-tentively looks firstly at his mirror image, but then also at his behind and his crust of bread in a mirror . . . obviously understanding the use of a mirror."[31]

Similarly, the German gestalt psychologist Wolfgang Köhler in 1925 commented on the lasting interest of chimpanzees in their mirror image; they continue to play with it, making strange faces at themselves and checking reflected objects against the real thing by looking back and forth between the two. Monkeys, in contrast, react with facial expressions that are anything but frivolous: they regard their reflection as another individual, treating it as a stranger of their own sex and species.

Compelling evidence was derived in the 1970s from elegant experiments by Gordon Gallup, an American comparative psychologist. An individual unknowingly received a dot of paint in a specific place, such as above the eyebrow, invisible without a mirror. Guided by their reflection, chimpanzees and orangutans—as well as children more than eighteen months of age—rubbed the painted spot with their hand and inspected the fingers that had touched it, recognizing that the coloring on the reflected image was on their own face. Other primates—and younger children—failed to make this connection. Gallup went on to equate self-recognition with self-awareness, and this in turn with a multitude of sophisticated mental abilities. The list encompassed attribution of intention to others, intentional deception, reconciliation, and empathy. Accordingly, humans and apes have entered a cognitive domain that sets them apart from all other forms of life.[32]

Nevertheless, sharp dividing lines—regardless of whether they place our species in a class by itself or create a slightly more inclusive cognitive elite—are to be treated with reservation. The mirror test provides a rather narrow measure of self-awareness. After all, such awareness may express itself in myriad other kinds of behavior and involve senses other than the visual. What to think of the dog's olfactory distinction between his own urine markings and those of other dogs; the bat's ability to pick out the echoes of its own sounds from among those of other bats, and the monkey's perfect sense of how far its hands, feet, and tail can reach in the surrounding canopy?

Some cognitive psychologists view the self as the interface between an organism and its environment. According to J. J. Gibson, the more complex an organism's interactions with its environment, the better it needs to know itself. This premise applies to the physical environment, but perhaps even more to the social milieu. A macaque or baboon can hardly function without knowing the social position of each group mate, the kinship network, which individuals are likely to side with each other in a fight, the possible reactions of others to particular actions, and so on. How could a monkey ever reach such a grasp of social affairs without knowing its own capacities and limitations, and its own position vis-à-vis others? Understanding one's surroundings equals understanding oneself. In this broader view, some species may reach greater heights of self-knowledge than others, but surely there can be no species without any such knowledge at all.

Similarly, it is hard to imagine empathy as an all-or-nothing phenomenon. Many forms of empathy exist intermediate between the extremes of mere agitation at the distress of others and a full understanding of their predicament. At one end of the spectrum, rhesus infants get upset and seek contact with one another as soon as one of them screams. At the other end, a chimpanzee recalls a wound he has inflicted, and returns to the victim to inspect it.

Human ethics everywhere urge us to adopt someone else's perspective and look at the world through the eyes of others, as in the Golden Rule: *do unto others as you would have them do unto you*. Perhaps the evolution of role-taking, which is a very special capacity indeed, began with rather simple forms. For example, monkeys seem perfectly capable of identification with another monkey. If Azalea's sister interrupts hair plucking of her retarded sibling by another monkey, even though Azalea herself does nothing to draw attention, or if a mother monkey hurries over to stop an approach by her child toward an ill-tempered individual well before anything happens, these actions suggest great sensitivity to potential harm involving others.

It is not hard to see why monkeys would want to avoid harm to themselves, but why would harm to another bother them? Probably they see certain others as extensions of themselves, and the distress of those resonates within them. Known as *emotional contagion*,[33] this mechanism initially operates indiscriminately, yet becomes more selective with age. Monkeys learn to recognize subtle signs of distress, even situations in which distress is merely imminent. They follow closely what happens around them, especially if it involves friends and relatives.

Full-blown role-taking involves quite a bit more, however. The other is recognized not just as an extension of the self, but as a separate entity. Cognitive empathy is the ability to put oneself in the "shoes" of this other entity without losing the distinction between self and other. The American psychologist Martin Hoffman believes that this remarkable capacity grows out of emotional contagion. Being vicariously affected by others may make the child curious about their internal state, and stimulate him to search for cues about the others' feelings. Out of this challenge grows an increased awareness of the self in relation to others.

The same challenge may have occurred in the course of evolution. Perhaps some species evolved social organizations in which it became particularly advantageous to appreciate how companions were doing—not just at an emotional level, but also by imagining their situation. Sharper awareness of the other entailed increased self-awareness. If the mirror test somehow taps into this ability, as Gallup suggests, higher levels of empathy may be limited to humans and apes.

One indication of a relation between the two is that the first signs of cognitive empathy in children appear at about the same time as mirror self-recognition.[34] Another sign is that consolation occurs in a self-recognizing species, the chimpanzee, but apparently not in macaques. Do macaques rarely reassure victims of aggression because they lack the required ability to trade places mentally with them? Signs of distress in others do affect them, but once the fight is over and these indications subside, macaques quickly lose interest.[35]

As so often with regard to gradual processes, the tension between continuity and discontinuity cannot easily be resolved. Even if temperatures change steadily, there is an abrupt change in properties when water turns into ice or steam. Both the gradualist and the believer in fundamental distinctions have a point when considering the evolution of empathy and sympathy. Yes, apes do share our capability for self-recognition in a mirror, but no, this capability does not necessarily mean that humans and apes are the only animals conscious of themselves. And yes, apes do show remarkable empathy, but no, they are not the only animals sensitive to the needs of others. We only need think of the incredible assistance elephants, dolphins, and lemurs offer each other to realize how widespread and well developed these tendencies are. Caring responses go back much further in evolutionary history than the ape-human lineage.

In the last few years, self-awareness and mirror tests have become hot topics of debate. The controversy will no doubt fertilize large areas of the behavioral sciences, as it touches on many issues. Apes can be expected to continue to steal the spotlight, for they may be the only animals other than ourselves to care about how they look in the eyes of others. Apes faced with a mirror inspect hard-to-see body parts such as their teeth and posterior. Female chimpanzees twist their bodies to get a good look at the pink genital swellings that arouse the males. Orangutans place vegetables on their heads to size up the effect. Even without mirrors, apes adorn themselves—albeit with peculiar taste. A chimpanzee will find a dead mouse and carefully place it between her shoulders for the entire day, taking care that it does not fall off, or drape vines around her neck in an act of self-beautification.[36]

Undoubtedly, the interest apes show in themselves relates to the complexity of their social life, in which it probably matters greatly how one is perceived. Like the Biami, apes do not need reflective surfaces to gain self-awareness. They are used to watching themselves in the *social* mirror: the spectators' eyes.

Lying and Aping Apes

A female guppy courted by two males ends up associating with one of them while another female follows the entire process from an adjacent tank. When this guppy "voyeuse" is introduced to the same males to see which one she likes better, she follows her predecessor's choice. Lee Dugatkin, an American ethologist who conducted these experiments, speculates that female guppies rely on each other's assessments of potential mates. The I-want-what-she-wants principle that Dugatkin found had the power of reversing a female's independent preferences known from earlier tests.

Similarly, two Italian scientists, Graziano Fiorito and Pietro Scotto, trained an octopus to attack either a red or a white ball. After the training another octopus was allowed to watch four demonstrations from an adjoining tank. The spectator closely monitored the actions of the demonstrator with head and eye movements. When the same balls were dropped in the spectator's tank, he attacked the ball of the same color as the first octopus.

What both experiments teach us is that even animals with minuscule brains compared to primates notice how members of their own species relate to the environment. The octopus identified with the other octopus and the female guppy with the other female guppy, both letting their counterpart influence their attitude toward a stimulus.

If *identification* is the ability to feel closer to one object in the environment than another, and to make the situation of the first to some extent one's own, this is a very basic ability indeed. It makes it possible to reach out mentally to others, making them an extension of the self, paying close attention to their situation so as to influence it or gain information from it. Identification underlies both empathy and imitation. The precision with which one individual can copy the behavior of another depends on the degree to which that individual is able to assume the other's point of view (another way of saying that the level of imitation depends on the level of empathy).

The simplest form of imitation is mere behavioral copying without realization of the benefit of the behavior. This is perhaps what guppies and octopi do, and what primates do much of the time as well. The premier copier is no doubt the chimpanzee.[37] Juveniles in the Arnhem Zoo would amuse themselves by walking single file behind a female named Krom, which means "crooked," all with the same pathetic carriage. They also walked around supporting themselves on both wrists—instead of on their knuckles as any self-respecting knuckle-walker is supposed to do—resulting in awkward locomotion similar to that of an adult male in the group whose fingers had been mangled in a fight.

Captive chimpanzees, furthermore, learn from watching people how to use tools such as hammers, screwdrivers, and brooms. That they do not always grasp the utility of the tool was noted by pioneering field-worker Robert Garner as long ago as 1896. Given a saw, his chimpanzee "applied the back of it, because the teeth were too rough, but he gave it the motion . . . He would put the back of it across a stick and saw with the energy of a man on a big salary."[38]

It is widely assumed that primates excel at imitation—so much so that we call it *aping*. Usually we mean more than mere copying. Experts on primate behavior are not in agreement about the more advanced kinds of imitation, however. In its most complete form the imitator adopts the model's perspective and recognizes both the model's goal and his method of bringing this goal closer. Are monkeys or apes aware of the problems others face? Do they understand others' solutions, and can they then apply this knowledge to the same problems? Whereas there is little or no evidence that monkeys do so, some scientists believe apes to be different.

The various technologies of wild chimpanzees, such as nut-cracking with stones or termite fishing with twigs, require fine manual skills that take years to acquire (and adult chimpanzees are said to be far better at them than naive humans). Young chimpanzees seem to watch closely and learn from adults. There are even stories of mothers' correcting their offspring's mistakes, which would amount to active teaching. Unfortunately, these observations are made under uncontrolled conditions, and field-workers have a rather fragmentary picture of their subjects' learning histories. Experimental psychologists who have carefully tested captive chimpanzees are not convinced that such high-level processes take place. They do agree that chimpanzees pick up information from watching others (such as the location where rewards are to be obtained, and the kind of tools that produce them), and that this *helps* them find a solution; yet they believe that ultimately problems are solved by each individual independently.[39]

The imitation controversy pertains to the broader question of whether animals view each other as having intentions, feelings, beliefs, and knowledge. Do they look at each other as sentient beings? This question bears directly on ethics: perceived intentions are the stuff of moral judgment. In our daily lives it matters greatly whether someone hurts us deliberately or accidentally. As soon as children can produce sentences, they begin arguing along these lines, confronting their parents with the difficult decision between the wet child's "He sprayed me on purpose!" and the dry child's "I didn't know he was there!" With praise and blame being meted out on the basis of our reading of other people's intentions, it is important to know if animals recognize knowledge or intention behind the behavior of others.

Pioneering research in this area was conducted in the 1970s by an American experimental psychologist, Emil Menzel, with nine juvenile chimpanzees. He would take one of them out into a large enclosure to show hidden food, or a frightening object such as a stuffed snake or crocodile. Subsequently, he would take the "knower" back to the waiting group and release all of them together. Would the others appreciate the knower's information, and if so, would they develop strategies to exploit it? Would the knower develop counterstrategies? All such behavior might require that chimpanzees know about knowing. Few accounts are more emblematic of the modern approach to social behavior than the following one by Menzel. He describes the attempt of a dominant chimpanzee, Rock, to extrapolate the knowledge of an equally canny subordinate, Belle, and outwit her.

If Rock was not present, Belle invariably led the group to food and nearly everybody got some. In tests conducted when Rock was present, however, Belle became increasingly slower in her approach to the food. The reason was not hard to detect. As soon as Belle uncovered the food, Rock raced over, kicked or bit her, and took it all.

Belle accordingly stopped uncovering the food if Rock was close. She sat on it until Rock left. Rock, however, soon learned this, and when she sat on one place for more than a few seconds, he came over, shoved her aside, searched her sitting place and got the food.

Belle next stopped going all the way. Rock, however, countered by steadily expanding the area of his search through the grass near where Belle sat. Eventually, Belle sat farther and farther away, waiting until Rock looked in the opposite direction before she moved toward the food at all—and Rock in turn seemed to look away until Belle started to move somewhere. On some occasions Rock started to wander off, only to wheel around suddenly precisely as Belle was about to uncover the food.

In other trials when we hid an extra piece of food about 10 feet away from the large pile, Belle led Rock to the single piece, and while he took it she raced for the pile. When Rock started to ignore the single piece of food to keep his watch on Belle, Belle had temper tantrums.[40]

Could Rock have been so persistent without recognition on his part—conviction, really—that Belle knew something that she did not wish to reveal? This explanation certainly would seem the most tempting, although simpler accounts, based on quick learning and anticipation of the other's actions, cannot be ruled out entirely.

In recent years research on attribution and perspective taking has exploded into experiments on both children and apes. Julie Hadin and Josef Perner showed children a picture series in which a girl puts an escaped rabbit back into its cage, then leaves, after which a boy removes the same rabbit from the cage to take it home. The children are then asked where the girl thinks the pet is, and whether she will be surprised to find it at the boy's home. In order to predict her reaction correctly, the children need to understand that the girl's knowledge and their own knowledge of the whole situation are different. Children begin to make such distinctions only by the age of six years.

Daniel Povinelli tested the attributional capacities of chimpanzees in the laboratory of Sarah Boysen. The location of food was pointed out to the apes by either a person who could have seen where the food was hidden, or by one who lacked this knowledge because his or her head had been covered at the critical moment by a paper bag (evidently, nobody worries about our species' image in the eyes of animal subjects!). Because the chimpanzees reacted differently to the advice from these two sources, they seemed to realize that knowing requires seeing.[41]

Monkeys may not recognize this distinction. Using the tendency of macaques to vocally "announce" food, Dorothy Cheney and Robert Seyfarth measured their perspective-taking abilities. A caretaker placed apple slices in a food bin in full view of a mother and her juvenile offspring. In another arrangement only the mother could see the food arrive because her offspring, although nearby, was hidden behind a partition. Mothers did not call more under the second condition, as they should have had they been sensitive to their offspring's point of view and lack of information.

The difference between chimpanzees and macaques emerging from these experiments has been corroborated by Povinelli in a telling study of role-taking. A chimpanzee was taught to select one of four handles. If she pulled the correct one, both she and a human at the other end of the apparatus would obtain food. The human but not the chimpanzee could see which handle was baited, and the human would point at this handle to assist the chimpanzee. It was a happy arrangement, and the chimpanzee soon learned to act according to the hints of her partner. After a large number of trials the roles were suddenly reversed, with the human now pulling the handles and the chimpanzee seeing the hidden food. Three of the four chimpanzees understood what was expected of them, having grasped the nature of the informant role from mere watching: they began helping their partner select the right lever. Yet when rhesus monkeys were allowed to work with human informants, none of them responded with the same sort of immediate understanding when the tables were turned: they first had to *learn* the new contingencies. It may be that chimpanzees can picture themselves in someone else's position and adopt this individual's role, whereas monkeys cannot.

The final, but perhaps least investigated, sign of higher cognition in chimpanzees is their deceptive nature. True deception—one of those capacities that we employ all the time without taking too much pride in it—can be defined as the deliberate projection, to one's own advantage, of a false image of past behavior, knowledge, or intention. In its most complete sense, it requires awareness of how one's actions come across and what the outside world is likely to read into them.

Chimpanzees may possess such awareness; their deceitful tactics have long been known by people who have raised them at home or worked with them in captivity. Many chimpanzees will, for example,

Instances of spontaneous deception in the Arnhem chimpanzee colony.[42]

Example 1.   A chimpanzee's teeth-baring signals nervousness. A grin on a male's face may therefore undermine the effectiveness of his intimidation display.

"The most dramatic instance of self-correction occurred when a male, who was sitting with his back to his challenger, showed a grin upon hearing hooting sounds. He quickly used his fingers to push his lips back over his teeth again. This manipulation occurred three times before the grin ceased to appear. Only then did the male turn around to bluff back at his rival."

Example 2.   Among thousands of records of reconciliation, six mention a dramatic shift from friendly to aggressive behavior. All instances followed unsuccessful attempts by an older aggressor to catch a younger victim, and involved unusually harsh punishment once the victim had come within reach.

"Puist, a masculine-looking female, aggressively pursues and almost catches a younger female. The victim screams for a while after her narrow escape, then sits down, panting heavily, and rests. The incident seems forgotten, yet approximately ten minutes later Puist approaches and makes a friendly gesture from a distance: she reaches with an open hand toward the other. The younger female hesitates, then approaches Puist with all the signs of mistrust—frequent stopping, looking around, and a slight grin on her face. Puist persists in her friendly invitation, adding soft pants when the younger female comes close. Soft pants have a particularly friendly meaning; they are often followed by a kiss, the chimpanzee's conciliatory gesture par excellence. Suddenly Puist lunges and grabs the younger female, biting her fiercely before she can free herself."

Example 3.   Dandy, the youngest adult male, did not always gain access to food if housed at night with the other adult males. The others would threaten him away.

"After a few months the keeper reported that in the twenty minutes or so between entering the cage and feeding time, Dandy was always unusually playful and often engaged the whole male band in play. Arriving with food, the keeper would find them romping around, piling straw on each other and 'laughing' [hoarse guttural sounds associated with play]. In this relaxed atmosphere Dandy would eat undisturbed, side by side with the others. Apparently Dandy feigned a good mood in order to influence the mood of others to his own advantage."

Example 4.   Low-ranking males mate with females at their own peril, as dominant males tend to interrupt such sexual activities. As a consequence, the lower-ranked males often hide in a secret rendez-vous that requires female cooperation.

"This kind of furtive mating is frequently associated with signal suppression and concealment. I can remember the first time I noticed it very vividly indeed, because it was such a comical sight. Dandy and a female were courting each other surreptitiously. Dandy began to make advances to the female, at the same time restlessly looking around to see if any of the other males were watching. Male chimpanzees start their advances by sitting with their legs wide apart revealing their erection. Precisely at the point when Dandy was exhibiting his sexual urge in this way, Luit, one of the older males, unexpectedly came around the corner. Dandy immediately dropped his hands over his penis, concealing it from view."

---

quickly collect a mouthful of water from a faucet in their cage when they see a stranger approaching, then wait with a perfect poker face until they can let the intruder have it. Some are such experts that they trick even people who are thoroughly mindful of the possibility. The ape will stroll around in his cage as if occupied with something else, only to swing around at the right moment when he hears his victim behind him.

The first body of evidence that chimpanzees do the same to one another was put together in 1982 in *Chimpanzee Politics,* based on my observations of the Arnhem colony. With years of macaque watching behind me before I was introduced to chimpanzees, I was totally unprepared for the finesse with which these apes con each other. I saw them wipe undesirable expressions off their face, hide compromising body parts behind their hands, and act totally blind and deaf when another tested their nerves with a noisy intimidation display. It is not hard to see how concern about signals emitted by one's own body relates to self-awareness—these chimpanzees acted quite differently from the puppy chasing its own tail.

What is special about lies is that they rapidly lose effectiveness if employed too frequently. For this reason striking instances of deception can be expected to be rare, and anecdotes are probably all we will ever obtain. Work in this area has been criticized yet there is nothing wrong with unique observations: the fact that human moon landings have been rare does not make their occurrence dubious.

Menzel implied the same when asking: "By the way, does anyone have any experimental evidence for deceit in any human president, king, or dictator? All I have ever seen here is anecdotes."[43]

Admittedly, there is an element of judgment, hence risk of over-interpretation, in anecdotal reports. Skeptics would need to explain, however, why chimpanzees appear to lend themselves more readily than other species to interpretations of deception. Does this fact tell us something about the apes or about the scientists attracted to them? The same interpretations simply do not present themselves with monkeys, as I noticed when I returned to watching macaques—over a hundred of them for a decade—and tried to collect instances of deception. The observer was the same, the desire to interpret (or over-interpret) behavior was undiminished, yet nothing even remotely comparable to what chimpanzees do was to be seen.

Similarly, when two British primatologists, Andrew Whiten and Richard Byrne, invited field-workers and other experts on primate behavior from all over the world to send them instances of spontaneous deception, they compiled 253 accounts, the most striking and complex of which concerned chimpanzees.

Even though further research is obviously needed, reports of special abilities in apes—mirror self-recognition, imitation, expressions of empathy, and intentional deception—are varied enough to suggest that somewhere in evolutionary history an important step was taken. Every piece of evidence in and of itself may not convince, but the simultaneous appearance of a number of indications of higher cognition in a single evolutionary branch is hard to explain without assuming an underlying principle that is absent or less developed in other branches.

Simian Sympathy

Cognitive evolution does not invent new categories of behavior. It works with, rather than replaces, the ancient emotional infrastructure, transforming it by an ever-greater understanding on the part of the actors. As a species, however, we value intelligence so highly that we tend to think that our reasoning capacities *drive* behavior. We are so skilled at producing convincing rationalizations that we begin to believe in them: the myth of rational man, or woman.

An example of overestimation of our rational nature is the "decision" to carry babies on the left hip. Right-handed mothers point out that it makes perfect sense to do so because they need their right hand to stir pots and fold laundry. Left-handed mothers have an equally compelling reason: obviously, a baby is more securely held in the dominant extremity. The fact is that a large majority of women show a left-side cradling bias regardless of handedness or cultural background. Men do not show the bias—they tend to carry babies on their right arm—but the left bias has also been reported for ape mothers. Because it transcends handedness, cultural, and even species bariers, it is unlikely that the left cradling bias has much to do with the rationalizations offered. One theory is that it is a natural tendency that places the baby closer to the mother's heart.[44]

Similarly, if some scientists believe that empathy is based on language or that helping involves a rational weighing of costs and benefits, they are probably overestimating the power of human reasoning and underestimating the role of emotions and subconscious motivations. The word "empathy" was coined as a translation of *Einfühlung,* a German term that became popular in academic circles early in this century. Inasmuch as *Einfühlung* literally means "feeling into," in the sense of getting inside the feelings of someone else, the German word exquisitely balances the interpersonal and affective sides of the process. In a time in which empathy is discussed largely as a cognitive feat, and in which cognition is often compared to the working of a cold-blooded computer, we should remember that the human mind knows no neat dividing line between thought and feeling. One individual's caring for another depends on a *mosaic* of factors ranging from rational and cognitive to emotive and physiological.

For this reason the question at the heart of this chapter is not so much whether any creatures other than ourselves can feel sympathy based on empathy (too much an all-or-nothing phenomenon), but *which elements* of human sympathy are recognizable in other animals. Fellow-feeling doubtless exists among social animals; what we want to know is the degree to which it resembles that among people. The cognitive dimension of this problem has more to do with the precise channeling of succorant tendencies than with the tendencies themselves. As Philip Mercer, a British philosopher, put it: "There are appropriate and inappropriate ways of helping others, and it seems natural to suppose that the more fully we sympathize with someone the more likely we are to give him the kind of help he needs."[45]

Thus, in aiding a friend, I combine the helping tendency of cooperative animals with a typically human appreciation of my friend's feelings and needs. The forces that propel me into action are the same, but I carry out the mission like a smart missile instead of like a blind rocket. Cognitive empathy is goal directed; it allows me to fine-tune my help to my friend's specific requirements.

One element underlying all caring behavior is mutual attachment. We are not surprised to find that dolphins, elephants, canids, and most primates respond to each other's pain and distress, because the members of these species survive through cooperation in hunting and defense against enemies and predators. The evolution of succorant behavior and bonds of affection is understandable, as every individual's life counts for the rest of the group. In a given species, the urge to help others roughly matches the need for help. A tiger needs no assistance and has absolutely no urge to provide it to others, yet the human primate is rather powerless on his own and therefore enters into elaborate contracts of mutual assistance.

A connection between the level of mutual dependency and succorant behavior is confirmed by invalid care in yet another highly cooperative nonprimate, the dwarf mongoose. A British ethologist, Anne Rasa, followed the final days of a low-ranking adult male dying of chronic kidney disease. The male lived in a captive group consisting of a founding pair and its offspring. Two adjustments took place. First, the sick male was allowed to eat much earlier in the rank order than previously. Instead of being chased away, he could nibble on the same piece of food as his father, the alpha male. Second, the rest of the group changed from sleeping on elevated objects, such as boxes, to sleeping on the floor once the sick male had lost the ability to climb onto the boxes. They stayed in contact with him, grooming him much more than usual. After the male's death, the group slept with the cadaver until its decay made removal necessary.

Following attachment, the next most basic ability involved in empathy is emotional contagion—vicarious arousal by the emotions of others. In its simplest form, there is total identification without discrimination between one's own feelings and those of the other. It is unlikely, for example, that a human newborn crying with the rest of the nursery has any notion that she is responding to feelings that originated with someone else. Rather, she seems hooked up to a communication network of direct lines between individual experiences without relay stations to tell her where the calls originated. Infants seem to "lose" themselves in collectivities of agony, joy, and sleepiness.

Caring behavior, even in the largest-brained species, may further

Increased awareness of self and other →

	Emotional contagion	Cognitive empathy
Attachment		Perspective shift
	Learned adjustment	Intentional deception
Identification		Attribution
	Behavioral copying	True imitation

Cooperative animals with strong attachments identify with one another and are sensitive to the other's emotions. To be genuinely concerned about another's well-being, however, requires differentiation between self and other. Most likely this cognitive step was taken by the ancestors of humans and apes, although its occurrence in other social animals cannot be excluded.

depend on innate releasers, that is, particular stimuli that almost automatically trigger a response in all members of the species. If I approach a chimpanzee friend with a scab on my arm, her eyes will brighten and focus on my little injury. She will beg for access, and, if I would let her, will open up the wound, all the while tooth-clacking with excitement. I could provide her no greater joy. And if one believes that the notion of "innate releasers" does not apply to humans, think again: I know people with exactly the same skin-peeling itch!

Another standard cue is known as the *Kindchenschema*: infantile traits, such as large eyes and rounded features, melt our hearts as well as those of other animals. No one better understood this than Walt Disney. The result was "cutification" of animals on a massive scale. Add dependency indicators, such as lassitude and a high-pitched voice, and one has the perfect formula to arouse protection and care, and to inhibit behavior normally reserved for older juveniles, such as the rejection of contact or punishment of misconduct. The special treatment enjoyed by Azalea and Wania-6672 may have been based on such characteristics; that is, other group members may never have reached the point of seeing them as the juveniles they were because they retained their babyish appeal.

An example of the prolongation of care in the wild is a capuchin monkey born in the Venezuelan jungle with partially paralyzed legs. The monkey could climb but not jump, and needed to be carried in order to get from tree to tree. According to John Robinson, the group carried the infant more than usual for its age (in capuchins many individuals other than the mother transport infants). The bad news was that the infant consumed normal amounts of food. Having no exercise, it grew big and fat and became more and more of a burden. The group bravely carried this butterball until it was seventeen months old, when it disappeared. No one knows how its life ended; it might have been abandoned by the group or snatched by a bird of prey.

Attachment, emotional identification, and innate responses, combined with potent learning abilities, provide a firm enough basis for elaborate caring behavior, which may sometimes be hard to distinguish from human expressions of sympathy. The latter differ, however, in that we recognize the other's experiences as belonging to the other, which is the only way we can feel genuine concern. A mother who shuts her eyes and grimaces when the doctor is about to stick a needle in her child's arm is anticipating the emotional disturbance of her child, hence of herself, while knowing full well that it is the child, not she, who will feel the pain. Identifying with and caring about another without losing one's own identity is the crux of human sympathy. As we have seen, this requires certain cognitive abilities, the most important one being a well-developed sense of self and the ability to assume another individual's perspective.

Two examples of simian sympathy illustrate the advantage of this ability. They describe succorant behavior of greater complexity and solicitude than found outside the human-ape branch of the primate tree.

The moat and the chain   The bonobos at the San Diego Zoo used to live in a grotto-type enclosure separated from the public by a 2-meter-deep dry moat. The moat was accessible to the apes by a chain hanging down into it; they could freely descend and climb up again. In *Peacemaking among Primates,* I described a situation repeatedly observed if the dominant male, Vernon, disappeared into the moat. A younger male, Kalind, would quickly pull up the chain and look down at Vernon with an open-mouthed play face—the ape equivalent of laughing—while slapping the side of the moat. On several occasions the only other adult, Loretta, rushed to the scene to "rescue" her mate by dropping the chain back down and standing guard until Vernon had gotten out. Both Kalind and Loretta seemed to know what purpose the chain served for someone at the bottom of the moat and acted accordingly—the one by teasing, the other by assisting.

Tired of the tire problem   The Arnhem chimpanzees spend the winters indoors. Each morning, after cleaning the hall and before releasing the colony, the keeper hoses out all the rubber tires in the enclosure and hangs them one by one on a horizontal log extending from the climbing frame. One day Krom was interested in a tire in which the water had been retained. Unfortunately, this particular tire was at the end of the row, with six or more heavy tires hanging in front of it. Krom pulled and pulled at the one she wanted but could not move it off the log. She pushed the tire backward, but there it hit the climbing frame and could not be removed either. Krom worked in vain on this problem for over ten minutes, ignored by everyone except Otto Adang, my successor in Arnhem, and Jakie, a seven-year-old male chimpanzee to whom Krom used to be the "aunt" (a caretaker other than the mother) when he was younger.

Immediately after Krom gave up and walked away from the scene, Jakie approached. Without hesitation he pushed the tires off the log one by one, as any sensible chimpanzee would, beginning with the front one, followed by the second in the row, and so on. When he reached the last tire, he carefully removed it so that no water was lost and carried the tire straight to his aunt, where he placed it upright in front of her. Krom accepted his present without any special acknowledgment and was already scooping water with her hand when Jakie left.

Inasmuch as cooperation is widespread in the animal kingdom, the tendency to assist a member of one's own species is nothing new or original. Yet the precise *intention* behind it changes as soon as the actor can picture what his assistance means to the other. It is hard to account for Jakie's behavior without assuming that he understood what Krom was after and wished to help her by fetching the tire. Perspective-taking revolutionizes helpful behavior, turning it into *cognitive altruism,* that is, altruism with the other's interests explicitly in mind.

A World without Compassion

When the chimpanzees in Gombe National Park suffered an epidemic of poliomyelitis, probably contracted from people, the partially paralyzed victims were treated with fear, indifference, and hostility, as if they had ceased to belong to the community. According to Jane

Goodall, Pepe met with the following reaction when he first showed up at camp, his useless arm trailing in the dust.

[The others] stared for a moment and then, with wide grins of fear, rushed for reassurance to embrace and pat each other, still staring at the unfortunate cripple. Pepe, who obviously had no idea that he himself was the object of their fear, showed an even wider grin of fright as he repeatedly turned to look over his shoulder along the path behind him.[46]

At first, healthy individuals avoided the polio victims because of their odd movements and dragging limbs. Next they directed charging displays at them; then they attacked.[47] Shunning grotesquely deformed individuals may be psychologically understandable (as well as adaptive, given the risk of infection), but it is one of many examples of the absence of compassion and mercy in a species in which at other times these very same qualities seem incipient. Brutal intercommunity violence observed between chimpanzees who in earlier years had peacefully traveled and groomed together, and the chimpanzee habit of tearing limbs and meat from prey animals who are still very much alive, such as a screaming colobus monkey, indicate that whatever inhibitions and sensitivities these apes possess, they are easily overruled by other interests.

The attitude toward members of one's own species should of course not be equated with that toward other species. Lack of concern for other species is to be expected, given the virtual absence of attachment. Animals often seem to regard those who belong to another kind as merely ambulant objects. Sue Boinsky reports that when an angry capuchin male in the wild ran out of ammunition while hurling things at her, he simply turned around, grabbed an unsuspecting squirrel monkey who sat nearby, and threw it at her as if it were just another branch. The capuchin, who would never have acted in this way with a member of his own species, clearly could not care less about the shrieking little monkeys with whom he shared the forest. Cruelty to other animals is something that we humans may have begun worrying about; it is a concern without precedent in nature. Hunters judge the hunted by caloric rather than emotional value, and even if other species are not perceived as food, usually nothing is to be gained by investing care in them.[48]

On rare occasions interspecific contact takes a cruel turn. Consider the game in which youngsters entice chickens behind a fence with bread crumbs. When they come within reach, the gullible chickens are hit with a stick or poked in the feathers with a sharp piece of wire. This "tantalus" game, in which the chickens were stupid enough to cooperate, was described by Wolfgang Köhler for his chimpanzees. They played the game for amusement, not to get hold of the fowl, and refined it to the point that one ape would be the baiter and another the hit man. It lends further support to the claim of higher cognition in chimpanzees: in the same way that empathy may lead to sympathy if combined with attachment, it may result in intentional harm if combined with indifference.

We are ready now for a second German term, *Schadenfreude,* which is quite the opposite of *Einfühlung* yet related to it for the same reason that sympathy and sadism are extremes that touch each other. Schadenfreude literally means "harm-joy." When people roll off their chairs laughing because Hardy falls into a barrel filled with glue while Laurel stupidly looks on, or because Chaplin gets slapped in the face by the object of his love, they are enjoying someone else's bad luck. The fact that so many comedians offer their audiences opportunities to express this type of glee indicates a deep-seated human urge to boost our self-esteem through the mishaps of others. We tend to conceal this emotion because in real life we experience it mainly in relation to people whom we do not like. The feeling probably derives from a sense of fairness: it is most predictably aroused if someone gets his comeuppance, as when a pompous or dishonest man loses his fortune. We do not feel Schadenfreude when the home of a poor family burns down, or when a child tumbles down the stairs, for these victims never threatened our self-worth.

Schadenfreude is the exact opposite of sympathy. Instead of sharing the pain of someone else, we get a kick out of it. Some of the most shocking, almost unbelievable, illustrations may be found in *The Mountain People.* Colin Turnbull describes an East African tribe, the Ik, who through starvation deteriorated to the point of dehumanization. All gaiety among the Ik seemed at the expense of others. They would shriek with laughter if someone fell, especially if weak or blind, or if an elder lost his food to roaming teenagers, who would go so far as to pry open the mouths of old people to pull out morsels that they had not yet swallowed. Harm to children was not exempted.

Men would watch a child with eager anticipation as it crawled towards the fire, then burst into gay and happy laughter as it plunged a skinny hand into the coals. Such times were the few times when parental affection showed itself: a mother would glow with pleasure to hear such joy occasioned by her offspring, and pull it tenderly out of the fire.

Turnbull never saw children more than three years old being fed by adults, and he observed that people ate away from their homes so as to avoid having to share the small quantities of food they had. Noting how pointless the quest for morality becomes under such inconceivable stress, and how nothing can be allowed to stand in the way of survival, the anthropologist cynically characterizes morality as "yet another luxury that we find convenient and agreeable and that has become conventional when we can afford it."[49]

What intrigues me most in this depressing account is that when love and sympathy are wiped out by circumstances, apparently it is not mere selfishness that raises its head but an actual delight in the misery of others. Could it be that people derive pleasure from an equalization of fate, regardless of whether it is brought about by an uplifting or downward movement? When we are well off, we root for the underdogs and like to see their lot improve. Yet when we are at the border of starvation ourselves, we are glad to see every bit of misfortune around us, as it confirms that we are not the only ones down and out.[50]

Whereas the possible link between sympathy and fairness has been little investigated, the connection with selfishness has received quite a bit of attention. Much of the philosophical literature, especially in the English-speaking world, has been marked by a juxtaposition of egoism and altruism. Rather than keeping alive the tension between these poles of sociality, there has been a tendency to come down on the side of the one or the other, usually the former. It must be obvious by now that I consider this paradigm rather sterile. The everything-is-egoism position is comparable to the claim that all life on earth is a conversion of sun energy. Both belong to the Great Truths of Science, yet in the same way that the second truth has never prevented us from recognizing the diversity of life, the first should not hold us back from making fundamental distinctions in the domain of motivation.

If I have a table full of food and you knock at my window, starving, I can invite you in and derive satisfaction from your happy face, or I can keep everything for myself and derive satisfaction from my own full belly. In both cases I may be called selfish; yet from your perspective, and that of society at large, it makes quite a difference which kind of self-interest I pursue. Also, the reward that I myself experience is altogether different in the two instances. Whereas most behavior is rewarded by how it affects the actor, acts of sympathy are rewarded by how the actor *imagines* that they affect the recipient. If this is selfishness, it happens to be the only kind that reaches out.

Experiments with human subjects by Robert Weiss and coworkers, in which the incentive for a response was the deliverance of another person from suffering, have confirmed that "the roots of altruistic behavior are so deep that people not only help others, but find it rewarding as well."[51] The fact that satisfying feelings tend to accompany acts of sympathy does not in any way detract from their other-directed nature if the only way to reap these rewards is via the other's well-being. When we calm a crying child, hugging and stroking him, we are not so much reassuring ourselves as we are the other. We monitor the impact of our behavior, and it fills us with immediate pleasure if the child laughs through his tears at a joke or a tickle. If human sympathy is indeed the "inborn and indestructible instinct" that David Hume, Arthur Schopenhauer, Adam Smith, and others, declared it to be, it is only natural that it comes with a built-in compensation in the same way that sex and eating do.[52]

With the feeling of sympathy so ingrained, it is only under the most extreme circumstances, as when people lose all means of subsistence or are packed together and starving in a concentration camp, that it dies. It becomes a thing of the past, a painful memory. An old Ik woman, to whom Turnbull had given some food, suddenly burst into tears because, as she said, it made her long for the "good old days" when people had been kind to one another. This common benevolence nourishes and guides all human morality. Aid to others in need would never be internalized as a duty without the fellow-feeling that drives people to take an interest in one another. Moral sentiments came first; moral principles, second.

Despite Immanuel Kant's opinion that kindness out of duty has greater moral worth than kindness out of temperament, if push comes to shove, sentiments win out. This is what the parable of the Good Samaritan is all about. A half-dead victim by the side of the road is ignored first by a priest, then by a Levite—both religious and ethically conscious men—yet receives care from the third passerby, a Samaritan. The biblical message is to be wary of ethics by the book rather than by the heart: only the Samaritan, a religious outcast, felt compassion.

One of my favorite experiments, by John Darley and Daniel Batson, re-created this situation with American seminary students. The students were sent to another building to give a talk about . . . the

Good Samaritan. While in transit they passed a slumped-over person planted in an alley. The groaning "victim" sat still with eyes closed and head down. Only 40 percent of the budding theologians asked what was wrong and offered assistance. Students who had been urged to make haste helped less than students who had been given lots of time. Indeed, some students hurrying to lecture on the quintessential helping story of our civilization literally stepped over the stranger in need, inadvertently confirming the point of the story.

Altruism is valued so highly precisely because it is costly and subject to all sorts of limitations and conditions. It may be subordinated to other interests or to felt obligations, and it may disappear entirely if unaffordable. In *The Moral Sense* philosopher James Q. Wilson characterizes sympathy as follows:

> It is easily aroused but quickly forgotten; when remembered but not acted upon, its failure to produce action is easily rationalized. We are softened by the sight of one hungry child, but hardened by the sight of thousands.[53]

Human sympathy is not unlimited. It is offered most readily to one's own family and clan, less readily to other members of the community, and most reluctantly, if at all, to outsiders. The same is true of the succorant behavior of animals. The two share not only a cognitive and emotional basis, therefore, but similar constraints on their expression.

Despite its fragility and selectivity, the capacity to care for others is the bedrock of our moral systems. It is the only capacity that does not snugly fit the hedonic cage in which philosophers, psychologists, and biologists have tried to lock the human spirit. One of the principal functions of morality seems to be to protect and nurture this caring capacity, to guide its growth and expand its reach, so that it can effectively balance other human tendencies that need little encouragement.

COGNITION AND EMPATHY

For primates, as for humans, to most effectively help others, one needs to understand their needs and feelings. To gauge the depth of animal empathy and social intelligence, studies focus on responses to distress, self-awareness, the transmission of information, and the manipulation of social relationships.

In an act of consolation, a juvenile chimpanzee rushes over to embrace Yeroen, who screams after having been defeated in a critical battle over leadership. *(From* Chimpanzee Politics; *Arnhem Zoo)*

Learned adjustment in the context of play. When partners differ in size and strength, as do the adolescent male bonobo and infant here, the older partner needs to restrain himself to keep the game fun for the younger. *(San Diego Zoo)*

Kevin, an adolescent male bonobo, striking a philosophical pose. *(San Diego Zoo)*

A juvenile bonobo (right) showing concern for another. He grabs the arm of his buddy when the latter steps closer to the female on the left. The female presents a potential hazard, as she has been in a bad mood all morning and has chased the first male several times. *(San Diego Zoo)*

Mai (in center with back turned) gives birth while other chimpanzees gather around her. Atlanta, standing at her side, utters a loud scream when the baby drops into Mai's hands. Atlanta's reaction suggests identification with the mother. *(Yerkes Field Station)*

Two rhesus infants only a few months old. The female in front, who has just been molested by an adult, receives a hug from a male peer. In rhesus monkeys, such comforting behavior virtually disappears later in life. A rare exception is the case of Fawn (facing page), who huddles with an older sister (upper left) after having been attacked, while her mother (right) grooms one of the aggressors. *(Wisconsin Primate Center)*

The behavior of this macaque mother may reflect learned adjustment to the needs of her offspring. With an infant clinging to her back, the mother reaches for soybeans scattered into a warm-water spring in a Japanese nature reserve. Park officials say they need to keep first-time mothers out of the water because these females tend to drown their infants when diving for food. Experienced mothers are apparently more careful. *(Jigokudani Park, Japan)*

In studies of free-ranging Japanese macaques, Eishi Tokida presented them with a transparent tube containing a piece of apple. Only a few monkeys learned how to obtain the food with a stick or by rolling a rock through the tube. Other monkeys did not seem to learn from their example. The most successful individual, Tokei, occasionally used her own infants, pushing them into the pipe and retrieving them once they had put their teeth into the food. This photograph, however, shows Tokei (stick in hand) threatening her offspring away with a stare. Having grown too large for use as a live tool, the infant is now seen as competition. (*Jigokudani Park, Japan*)

Ropey, a rhesus monkey, holds her daughter (right) together with the alpha female's son. Perhaps such double-holding serves to promote friendship between the mother's own offspring and desirable peers. *(Wisconsin Primate Center)*

The opposite strategy is for a mother to interrupt contact: a high-ranking female breaks up the play between her daughter (left) and a low-ranking peer. She punishes the play partner, pushing her down roughly. *(Wisconsin Primate Center)*

Young chimpanzees test the social rules of their community by teasing their elders. A juvenile holds a stick behind his back, ready to throw it at a resting adult female. *(Arnhem Zoo)*

Faye, a young chimpanzee involved in studies of cognitive development conducted by Kim Bard at the Yerkes Primate Center, plays with a mirror. In an atypical facial expression, she moves her lower jaw up and down while staring straight at her reflection. It is as if she is exploring the connection between her own movements and those in the mirror. Such games are absent in most other animals, who either ignore their reflection or treat it as a stranger. Faye is close to the age, between twenty-eight and thirty months, at which most chimpanzees in Bard's research pass the mark test of mirror self-recognition.

Self-decoration by a juvenile bonobo, who has draped banana leaves around her shoulders. Afterward, she parades around with the leafy shawl. *(San Diego Zoo)*

Four juvenile chimpanzees nervously inspect a stuffed snake hidden in the grass of their enclosure during one of Emil Menzel's experiments on knowledge transmission. *(Courtesy of Emil Menzel; Tulane Primate Center)*

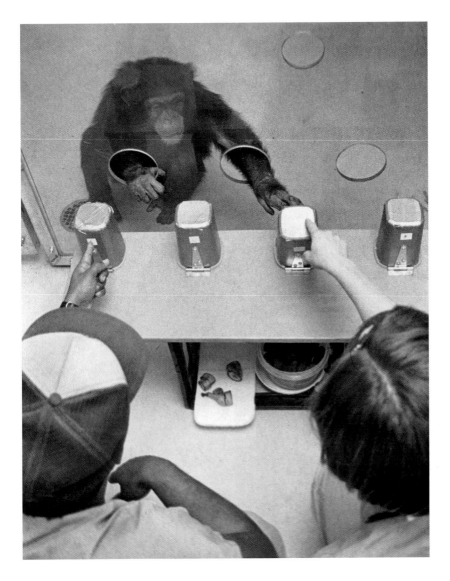

A chimpanzee receives advice from two experimenters about which cup covers food. Only one of the two persons knows; the other was absent or had occluded vision while the food was being hidden. The ape reaches for the cup pointed out by the experimenter on the right, who is indeed the one who knows. Possibly the subject understands that seeing leads to knowing. *(Photograph by Donna Bierschwale; courtesy of New Iberia Research Center)*

During hand-clasp grooming, the two partners hold hands above their heads, grooming each other with their free hand. Chimpanzees at the Yerkes Field Station spontaneously began using this posture a few years ago. Since then, it has spread within the group. The posture, never observed in other captive chimpanzees, is known in only two (geographically separate) wild communities. It thus seems to be a group-specific culturally transmitted behavior.

3

RANK AND ORDER

The overall picture of group organization in these animals [baboons] is of a sensitive balancing of forces, the balance being achieved by the social learning of individuals in the group from the time of birth to adulthood, so that infringements of the group norm are rare. When they do occur, they may be severely punished if the victim is caught.

Ronald Hall[1]

One balmy evening, when the keeper called the chimpanzees inside, two adolescent females refused to enter the building. The rule at Arnhem Zoo being that *none* of the apes will receive food until *all* of them have moved from the island into their sleeping quarters, the chimpanzees actively assist with the rule's enforcement: latecomers meet with a great deal of hostility from the hungry colony.

When the obstinate teenagers finally entered, more than two hours late, they were given a separate bedroom so as to prevent reprisals. This protected them only temporarily, however. The next morning, out on the island, the entire colony vented its frustration about the delayed meal by a mass pursuit ending in a physical beating of the culprits. Needless to say, they were the first to come in that evening.

A Sense of Social Regularity

That animals follow rules has been known for a long time. Female mammals, for example, threaten almost anyone or anything that approaches their young uninvited. They may do so in different ways or to different degrees, but maternal protection is widespread and highly predictable. So much so that we may declare it a rule—a *descriptive rule,* to be precise. Because this kind of rule describes typical behavior, it can be applied not only to animate objects, but to inanimate ones as well. For example, we can say that as a rule stones fall when released, whereas helium balloons do not.

Descriptive rules are not particularly interesting from a moral perspective, as they lack the crucial "ought" quality. Stones do not fall to avoid getting into trouble. Only animals and humans follow *prescriptive rules,* rules actively upheld through reward and punishment. With regard to other animals we notice this most readily if the rules are of our own design, such as those that we apply to pets and work animals. Yet the remarkable trainability of certain species, such as sheepdogs and Indian elephants, hints at the possibility of a rule-based order among these animals themselves.

To return to maternal protection, it is easy to see how it would affect the way others approach and treat the young of the species. In a chimpanzee colony, any individual deviating from the mother's standards will either meet with her wrath or find it more difficult in the future to obtain the infant from her. A prescriptive rule is born when members of the group learn to recognize the contingencies between their own behavior and that of the mother and act so as to minimize negative consequences. They learn how to handle the infant without making it scream; to bring it back of their own accord; not to climb to dangerous spots with the infant desperately hanging on, and so on. Such heedfulness is not well developed in juveniles, who still have much to learn about infant behavior and maternal reactions. Thus, when a juvenile carries an infant around, the mother is never far behind. When juveniles reach adolescence, most of them have learned the rules well enough that mothers trust them as baby-sitters.

In the stump-tailed macaque all group members, even the largest males, commonly avoid very young infants who wander about unattended. It is as if they are afraid of getting into trouble. This is also the only species of macaque in which infants have a distinct coat color—much lighter than that of older individuals. It makes them hard to miss in a monkey melee. As soon as a baby is near or on its mother, it attracts extraordinary attention. Mothers are typically surrounded by others who utter special vocalizations, known as staccato grunts, while trying to look closely at the infant's face or inspect its genitals. Do these grunts express endearment? Probably something other than that, because if love and affection were the main motivations we would expect mothers to grunt the most, yet this is not at all the case. For her dissertation research, Kim Bauers recorded hundreds of vocalizations in the stump-tail group at the Wisconsin Primate Center and found that whereas females commonly grunt at each other's babies, they *never* do so at their own offspring.

Bauers found that the shorter the distance between mother and infant, the more likely others will grunt when trying to contact the little one. Even though aimed at the infant, staccato grunts seem intended for the mother as well. Perhaps it is the way members of this species "ask permission" to approach an infant. The fact that mothers themselves do not need such permission would explain the absence of their grunts.

This interpretation is further supported by the appeasing effect of staccato grunts: silent interest in infants is far more often rebuffed by maternal threats and slaps than interest announced by a series of grunts. In other words, directing friendly sounds to an infant forestalls problems with its mother. This may be an acquired social convention; most silent approaches, hence most rebuffs, involve juveniles, who may still be in the process of learning how to overcome maternal protectiveness.

Of even greater interest than rules enforced by individual mothers are rules sanctioned by the community. In the case of the two latecomers in Arnhem, we saw a group response, but the regulation in question was of course put in place by people. Chimpanzees also seem to develop their own rules, however.

Jimoh, the current alpha male of the Yerkes Field Station group, once detected a secret mating between Socko, an adolescent male, and one of Jimoh's favorite females. Socko and the female had wisely disappeared from view, but Jimoh had gone looking for them. Normally, the old male would merely chase off the culprit, but for some reason—perhaps because the female had repeatedly refused to mate with Jimoh himself that day—he this time went full speed after Socko and did not give up. He chased him all around the enclosure—Socko screaming and defecating in fear, Jimoh intent on catching him.

Before he could accomplish his aim, several females close to the scene began to "woaow" bark. This indignant sound is used in protest against aggressors and intruders. At first the callers looked around to see how the rest of the group was reacting; but when others joined in, particularly the top-ranking female, the intensity of their calls quickly increased until literally everyone's voice was part of a deafening chorus. The scattered beginning almost gave the impression that the group was taking a vote. Once the protest had swelled to a chorus, Jimoh broke off his attack with a nervous grin on his face: he got the message. Had he failed to respond, there would no doubt have been concerted female action to end the disturbance.

These are the sorts of moments when we human observers feel most profoundly that there is some moral order upheld by the community. We cannot help but identify with a group that we watch day in and day out, and our own values of order and harmony are so similar that we would have barked along with the chimpanzees if we thought it would have mattered! Whereas some of us are inclined to explain the group's reaction to Jimoh in moral terms, such as "He just went too far," other observers might prefer a more neutral account along the lines of "Chimpanzees sometimes bark in response to aggression." There is one problem with the latter view, however: one never hears woaow barks when a mother punishes her own offspring, or when an adult male controls a tiff among juveniles—even if he uses force in the process. Not every fight triggers these calls. It is a reaction to a very particular kind of disturbance, one that seriously endangers relationships or lives. Thinking in terms of rules and violations may help us come to grips with its relevant features.

Undoubtedly, prescriptive rules and a sense of order derive from a hierarchical organization, one in which the subordinate pays close attention to the dominant. Not that every social rule is necessarily established through coercion and dominance, but prototype rule enforcement comes from above. Without agreement on rank and a certain respect for authority there can be no great sensitivity to social rules, as anyone who has tried to teach simple house rules to a cat will agree. Even if cat lovers fail to see a nonhierarchical nature as a shortcoming—on the contrary!—it does place their pets firmly outside the human moral realm. Evolved as solitary hunters, cats go their own way, indifferent to what the rest of the world thinks of them.

Respect for rules and norms can develop only when the opinions and reactions of others matter. Fear of punishment is important, but not the whole story: the desire to belong to a group, and to fit in, is also involved. According to Lawrence Kohlberg, who pioneered research in this field, these elements are recognizable in the first stages of human moral growth. Development begins with obedience and a wish to stay out of trouble, followed by an orientation toward approval and pleasing others. For the child, it is the adult's approval that is sought; for the adult, it may be that of an omnipotent God infused with absolute moral knowledge. There is obviously more to morality—Kohlberg's scheme counts six stages up to and including an autonomous conscience—yet submission to a higher authority is fundamental. This feature is also less peculiarly human than some of the abilities involved in the later stages: submission to authority is part of a primordial orientation found not only in our fellow primates, but in a host of other animals as well.

It cannot be accidental that obedience and a desire to please are conspicuous traits of man's best friend. That dogs provide almost a caricature of humanity's early moral stages may explain our species' love affair with the canine soul. Most of the time dogs are "good," otherwise we punish them for being "bad" in the hope of changing their behavior. At the same time that dogs are highly sensitive to praise and blame, however, the more advanced stages of human moral development, which emphasize rights and equality, are beyond their comprehension. They think in terms of vertical, not horizontal, arrangements. These animals do not take kindly, for example, to an antiauthoritarian upbringing. Unresolved status disputes provide much of the business of dog therapists. Owners who hate to be masters deprive their pets of the element they need most for psychological stability: a clearly defined social position. Many a dog who cannot be under his owner in the family pack will try to be on top—which, in turn, is bound to undermine the owner's psychological stability!

Dogs inherited their law-and-order mentality from pack-hunting ancestors. In much the way we teach a puppy behavioral rules, dogs and wolves seem to teach their young (see pages 94–95). A hierarchical orientation is by no means limited to the Canidae, however. It is also widespread in the primates, although in their case it is mitigated by a strong tendency to form alliances, that is, a tendency for two or more parties to band together against a third. Alliances usually bolster the position of dominants, but sometimes subordinates jointly stand against higher-ups. The resulting balancing of power, combined with a tendency for reciprocal exchange, produces the beginnings of an orientation to equity, particularly in the chimpanzee.

We can see this orientation when a group is faced with an attractive resource. Will the bosses claim everything for themselves, or will they share? In chimpanzees begging for food is common, and a beggar who is ignored may express frustration by throwing a fit. Temper tantrums are high drama, capable of inducing possessors to relinquish part of their food. The rhesus macaque, in contrast, lives in a society with rather intolerant dominants. From a safe distance the subordinate silently watches the dominant's food consumption. Sharing is absent

Canids have an excellent sense of social rules, which allows for order within the hunting pack and explains their trainability for human purposes. They not only follow rules, they may actively inculcate them in others. Below are examples in which dominant wolves or dogs seem to deliberately wait for or even induce a transgression in order to penalize it.[2]

Example 1. Eberhard Trumler describes how a father dog instills obedience, once his offspring have outgrown their puppy license.

"[One day] he 'declares' an old-bone taboo. First, the pups try to get hold of it. Immediately they are severely punished by the father, who grabs the violator by the scruff of his neck, shaking him vigorously. Naturally the victim yelps and, once released, throws himself submissively on his back. Shortly thereafter, however, when the father appears to be busy with something else, the disciplined pup circumspectly sneaks up to the forbidden bone again—only to receive another castigation. This may be repeated several times, and one gets the impression that the pups want to know exactly what kind of response they can expect from the old one. Anyone with a puppy at home will be familiar with such probing of the educator."

Example 2. Behavioral modification by a mother wolf is reported in *Of Wolves and Men,* by Barry Lopez.

"A female wolf left four or five pups alone in a rendezvous area in the Brooks Range one morning and set off down a trail away from them. When she was well out of sight, she turned around and lay flat in the path, watching her back trail. After a few moments, a pup who had left the rendezvous area trotted briskly over a rise in the trail and came face to face with her. She gave a low bark. He stopped short, looked about as though preoccupied with something else, then, with a dissembling air, began to edge back the way he had come. His mother escorted him to the rendezvous site and departed again. This time she didn't bother watching her back trail. Apparently the lesson had taken, for all the pups stayed put until she returned that evening."

Example 3. The American anthropologist and primatologist Barbara Smuts related another example of apparent rule-teaching seen in her dog, Safi, a mixed sheepherding breed. Safi is older than and dominant over a neighbor dog, an Airedale named Andy, with whom she plays every day in the yard of Smuts's house.

"I threw the ball for them repeatedly. Normally Safi always gets the ball; Andy defers to her even if the ball lands nearest to him. On this occasion the ball bounced in an odd way and landed right at Andy's feet while Safi stood at a distance. He grabbed the ball and brought it back to me. Safi followed with no sign of distress.

"I threw it again, and as usual Safi got it. But instead of returning it to me, as she normally does, she now brought the ball to Andy and dropped it right in front of him, backed off, and waited. Naturally Andy picked it up, at which point Safi pounced on him, pinned him to the ground, and held him with his neck in her mouth, growling softly. Andy immediately dropped the ball and was appropriately submissive. Safi released him and the game continued amicably—except that Andy hasn't been seen since going for the ball. It was as if Safi had intentionally communicated to him 'Thou shalt not pick up my ball!'"

Example 4. Early this century, Captain Max von Stephanitz, the greatest authority on the German shepherd dog, explained that there is a lot more to the learning of rules than fear of punishment. Note the distinctly moral perspective von Stephanitz takes when outlining the objectives of training this breed.

"If the dog is too keenly educated without love, then the young dog will be distressed of soul, and his capabilities will not be developed, because the foundation of confidence, which is cheerful trust, has failed. A healthy education will not make a shy broken slave without a will, or a machine working only when asked, but will make him submerge his own wishes to a higher judgment, and become a creature who works freely, and is pleased to be able to work. The education must awaken and develop dormant qualities and abilities, and moderate all excesses, strengthen all weaknesses, and guide all faults in the right direction."

---

in this species, as is begging and protest against monopolization. The contrast can be summarized by saying that rhesus monkeys have different *expectations* than chimpanzees about the distribution of resources. Chimpanzees count on a share; rhesus monkeys do not.

In analogy with the human sense of justice, we may call this a *sense of social regularity*, which I define as follows:

A set of expectations about the way in which oneself (or others) ought to be treated and how resources ought to be divided. Whenever reality deviates from these expectations to one's (or the other's) disadvantage, a negative reaction ensues, most commonly protest by subordinate individuals and punishment by dominant individuals.[3]

The sense of how others should or should not behave is essentially egocentric, although the interests of individuals close to the actor, especially kin, may be taken into account (hence the parenthetical inclusion of others). Note that the expectations have not been specified: they are species typical. Because the expectation, or at least the ideal, of equality is so pronounced in our own species, we perceive the rules among rhesus monkeys as less "fair" than those in our closest relative, the chimpanzee. More important than this human bias, however, is the fact that all species seem to act according to what they can (or have come to) expect from others, thus creating a stable and predictable modus vivendi among themselves.

An obvious problem for the ethologist—and one reason why these issues have not received the attention they deserve—is that expectations are not directly observable. Do animals even have them? We may empirically define an expectation as familiarity with a particular outcome to the degree that a different outcome has an unsettling effect, as reflected in confusion, surprise, or distress. Since O. L. Tinklepaugh's research in the 1920s, we know that a monkey who has learned to find a banana hidden in a particular location will act nonplussed if the banana has been secretly replaced by a mere leaf of lettuce. At first she will leave the lettuce untouched, look around, and inspect the location over and over. She may even turn to the experimenter and shriek at him. Only after a long delay will she "content" herself with the lettuce. How to explain such behavior except as the product of a mismatch between reality and expectation?

A second, even more intractable problem is that of intentionality. I speak without hesitation of social rules from the rule-follower's perspective. When a monkey learns that certain acts always provoke a negative reaction, he will begin to suppress these acts or show them with great circumspection; the monkey can then be said to have submitted to a socially enforced rule. But can we speak of rules from the implementer's perspective? Do animals deliberately teach one another how to behave, or do they just respond to particular situations with frustration, protest, and sometimes violence? I certainly do not wish to exclude the possibility that they purposely impose limits on the behavior of others and carefully monitor the slightest transgression in order to strengthen the rule (the previous boxed examples hint at this possibility in canids), yet I cannot say that the evidence is overwhelming. We do not know if the rules that we recognize in animal behavior, and that we see being enforced, exist *as rules* in the animals' heads. Without experimentation this thesis will be hard to prove. For the moment I qualify terms such as social "rule" and "norm" by adding that they refer to behavioral modification by others regardless of the intentionality of the process.

The only way to estimate a species' sense of social regularity is by paying as much attention to spontaneous social acts as to how these acts are *received* by others. We need to determine which kinds of behavior are accepted and which meet with resistance, protest, or punishment. Here is a brand-new research agenda, one that will reveal differences not only between species, but perhaps also between different groups of the same species.

It will be easy enough to find rules imposed from above, according to which dominants constrain the behavior of subordinates. From the perspective of morality, however, the exciting rules will be those that constrain the behavior of everyone, and emphasize sharing and reciprocity. Before exploring the various possibilities in greater depth, let me give a final example of a negative reaction to another individual's behavior. The incident, first reported in *Chimpanzee Politics*,[4] suggests that the chimpanzee's sense of social regularity is not concerned solely with hierarchical issues, but also with more advanced social arrangements, such as the familiar *One good turn deserves another.*

A high-ranking female, Puist, took the trouble and risk to help her male friend, Luit, chase off a rival, Nikkie. Nikkie, however, had a habit after major confrontations of singling out and cornering allies of his rivals, to punish them. This time Nikkie displayed at Puist shortly after he had been attacked. Puist turned to Luit, stretching out her hand in search of support. But Luit did not lift a finger to protect her. Immediately after Nikkie had left the scene, Puist turned on Luit, barking furiously. She chased him across the enclosure and even pummeled him.

If Puist's fury was in fact the result of Luit's failure to help her after she had helped him, the incident suggests that reciprocity in chimpanzees may be governed by obligations and expectations similar to those in humans.

The Monkey's Behind

Attributed to Saint Bonaventura, a thirteenth-century theologian, the saying "The higher a monkey climbs, the more you see of its behind" warns of the exposure of character flaws in people free of the usual social constraints. Power creates freedom of action, but often also brings vanity, mercurial mood swings, and constant worry over how long the power will be enjoyed. Reality may merge into fantasy for holders of absolute power; there is nothing to check their will. Few people have the self-discipline to handle this drug. Its trappings are partly due to the admiration and ingratiation received from others. If individuals cannot be powerful themselves, they seek the glow of someone else's power.

When the Great Chief enters a room, all heads turn and discussions come to a halt. We experience the presence of something larger than life. The chief may speak softly, yet everyone listens; he may tell a stale joke, yet everyone laughs; he may make an odd request, yet no one doubts that there must be an excellent reason. Why do we attribute superhuman qualities to such people, and why do we allow them to exploit our feelings of insecurity? Power is not an individual attribute; it is a relational one. For every powerful person there are others supporting that superiority, feeding that ego.

Like any pact, however, the one between dominant and subordinate is fragile. No one is more aware of this than the power holder himself. The more absolute the power, the greater the paranoia. Qin Shihuang, China's first almighty emperor, was so concerned about his safety that he hid all roads leading to his various palaces so that he could come and go unnoticed. More recently, it was discovered that Nicolea Ceausescu, the executed dictator of Rumania, had constructed three levels of labyrinthine tunnels, escape routes, and bunkers stocked with food beneath the Communist Party building on the Boulevard of Socialist Victory in Bucharest.

The vicious circle rotating between the hunger for power and the fear of losing it was summarized by Thomas Hobbes in *Leviathan:*

> I put for a generall inclination of all mankind, a perpetuall and restless desire of Power after power, that ceaseth onely in Death. And the cause of this, is not always that a man hopes for a more intensive delight, than he has already attained to; or that he cannot be content with a moderate power: but because he cannot assure the power and means to live well, which he hath present, without the acquisition of more.[5]

The desire to dictate the behavior of others is such a timeless and universal attribute of our species that it must rank with the sex drive, maternal instinct, and the will to survive in terms of the likelihood of its being part of our biological heritage. It may come as a surprise, therefore, that not all students of animal behavior believe in its existence. Animals do establish dominance orders, and do change ranks via tense confrontations, yet striving for the best positions is not always given the status of a distinct motivation. Some scientists would describe a series of rebellions leading to the overthrow of the alpha animal but painstakingly avoid any implication that the challenger might have set out to achieve this result. The process would be phrased in terms of action, reaction, and outcome, not intention. This view of animals as blind, ignorant actors in their own political dramas is as surprising to me as would be an Olympic athlete who represented her winning of the gold as something that had never even crossed her mind.[6]

The alternative view—at least as old if not older—is that animals deliberately strive to dominate others. In the 1930s Abraham Maslow, an American psychologist who later became famous for his theories about self-actualization, was one of the first to study social dominance in monkeys. He happened to do so in the same small zoo in Madison where I observed macaques decades later. Maslow described the cocky, confident air of dominant monkeys and the slinking cowardice, as he called it, of subordinates. He postulated a *drive for dominance,* in the same breath objecting to the term "submission" as it might imply that subordinates give up any hope of besting their superiors (which according to Maslow they never do). Despite this objection, he was one of the first to speculate about the function of submissive behavior, saying that it placates the dominant by admitting social inferiority.

Subordinate wolves greet those of higher rank by licking the corners of the superior's mouth (similar to the way dogs lick their owner's face), rhesus monkeys present their behind or bare their teeth in a grin, and chimpanzees and humans bow or prostrate themselves. While such displays attest to the extreme importance animals attach to social dominance, their significance was ignored during a period in which hierarchies were analyzed primarily in terms of who-gets-what. The result was such an obsession with the outcome of competition that other aspects fell by the wayside.

The first problem with this approach was that, in some species, who-gets-what is decided as much by social tolerance as by rank. Among chimpanzees, for example, it is not unusual for a female calmly to remove a male's food from his hands even if he tries to prevent deprivation by turning away. Does this interchange suddenly make her dominant over him even if there can be no doubt who would win a fight?

Second, confrontations are won and lost in many ways. To withdraw in a hurry is something quite different from an appeasing display: the first outcome breaks the relationship, the second tries to preserve it. If one sees an alpha wolf with erect, bristling tail surrounded by whining subordinates with their tails between their legs, or a swaggering alpha chimpanzee greeted by groveling subordinates who hurry over to him from great distances, it is evident that these encounters communicate much more than a mere win or loss. Status rituals reveal the deep structure of relationships and express cohesive tendencies as much as hierarchical ones. I will speak of *formal dominance,* to distinguish such external signs of status from everyday contests and their rather variable outcomes.[7]

Changes in formal rank often take place via a series of provocations by the previous subordinate, who may incur defeat and injury before winning any confrontations. A chimpanzee male who used to go out of his way to pay his respects to the boss, bowing all the way and nervously jumping away at the slightest threat gesture, is transformed into a defiant producer of noise and mayhem. He seems to have grown in size, displaying every day a little closer to the dominant, forcing him to pay attention by throwing branches and heavy rocks at him. In the beginning the outcome of these confrontations is rather open ended. Depending on the amount of support each rival receives from the rest of the group, a pattern will emerge, sealing the dominant's fate if it turns out to be in favor of his challenger. In all the processes that I have witnessed, the critical moment is not the first victory for the challenger, but the first time he elicits submission. The former dominant may lose numerous times, flee in panic, end up screaming high in a tree, and so on, but if he refuses to raise the white flag that the species evolved for this purpose, the challenger will not let up. Only when his target formally submits will the challenger change his conduct from aggressive to tolerant. The two rivals will reconcile, and calm will be restored.[8]

How to explain the incredible energy put into rank reversals, the life-endangering risks taken, and the abrupt change in attitude once the other submits, other than as an action-reaction chain *aimed at* forcing the other into recognizing a new order? I am a firm believer that primates (and many other animals as well) are aware of their dominance relationships and share our will to power in the sense that they actively try to improve their positions whenever shifts in alliances or physical abilities permit.

It is even possible that monkeys and apes are aware of more than just their own position relative to others. Robert Seyfarth has demonstrated that wild monkeys of adjacent rank groom one another more than those many positions apart, a finding confirmed by our own research. Seyfarth speculated that monkeys may be so familiar with their rank order that they not only realize who is above or below them, but also by approximately how many rungs on the ladder. This awareness would require an appreciation of rank relations among others.

Another possible indication of hierarchical knowledge is the so-called *double-hold*. Never reported by other students of primate behavor, it may be a local tradition found only (for reasons unknown) in the rhesus monkeys at the Wisconsin Primate Center. There, however, we have seen dozens of females do it hundreds of times.

Typically, a mother with her own infant attached to her belly picks up the straying infant of another female. She then holds the two infants together for a couple of minutes, with both arms wrapped around them as if she has twins, after which she releases the second infant. Obviously, this act by itself does not require any special knowledge. When we began paying more attention, however, we noticed that double-holding is highly selective: nine out of ten times mothers hold their infant with the offspring of females who outrank them. Since the second mother is usually not nearby, the implication is that female monkeys know for each infant in their group whether it is of high or low descent.

Given the general tendency of macaques to try to establish connections with higher-ups in the hierarchy, and given that these connections may pay off in terms of protection or tolerance around resources, the double-hold could be a way of starting this process early. In the same way that siblings develop ties because of their association with the same mother, infants may become close to anyone close to their mother, including peers of other families with whom they have been double-held. Perhaps mothers are "suggesting" upper-class rather than lower-class friends to their offspring. Not that I believe that rhesus monkeys have five-year plans for their young, but there could very well be a long-term benefit from this kind of behavior.

One might counter that it reflects nothing else than attraction to high-ranking infants, and that involvement of the mother's own infant is immaterial. I doubt, however, that this theory explains the high rate of double-holding. Furthermore, a few incidents suggest that the presence of the mother's own infant is not accidental. Our champion double-holder, Ropey, once was approached by an infant making one of his first excursions away from his mother, the alpha female. Ropey gave friendly lipsmacks to the infant, but did not pick him up. Instead, she glanced repeatedly at her youngest daughter, who was playing in another corner of the pen, meters away. Ropey then seemed to make a decision: she dashed away to collect her daughter, brought her back to the alpha's infant, and did a double-hold with the two of them.

The significance monkeys and apes attach to dominance relationships, and their jostling for positions and connections, mean that group life encompasses two conflicting strategies. The first is to probe the social order for weaknesses and look for openings to improve one's standing. Inasmuch as this strategy subverts existing structures and creates chaos, one might regard it as antisocial. Yet from the viewpoint of the parties knocking down the old walls, there is nothing antisocial about it; for them it is pure progress.

The second strategy is a response to the first: conservation of the status quo. Although very much in the interest of the parties with the best positions, the resulting stability also benefits the young and weak, who are the first to suffer in case of all-out war within a group. Hence the potential of a pact between top and bottom in which the lower echelons back the reigning powers, provided these guarantee their security.

Society results from the equilibrium between these contradictory strategies, so each society is more than the sum of its parts. However meticulously we were to test a single organism, we would learn little or nothing about the kind of organization to emerge from the interplay among many of the same organisms. No one doubts that such interplay has its own dynamics, yet much research still focuses on the individual at the expense of the system as a whole. This reductionist bias is revealed in our tendency (mine included) to speak of "dominants" and "subordinates" as if there are two kinds of individuals. Every member of society, however, except the ones at the very top and very bottom, represents both. As pointed out by the novelist and science philosopher Arthur Koestler, "The members of a hierarchy, like the Roman god Janus, all have two faces looking in opposite directions: the face turned towards the subordinate levels is that of a self-contained whole; the face turned upward towards the apex, that of a dependent part."[9]

For the same reason that the pattern of a sweater is lost once it has been pulled apart, social hierarchies cannot be understood by breaking them into their constituent parts. Occupants of top ranks do enjoy privileges—otherwise there would be no need to fight over the positions—yet by definition the top is a small portion of the hierarchy. Therefore, in addition to finding out what is in it for the "dominants," we need to know what is in it for the "subordinates" and consider the organization as a whole. All individuals are embedded in the same social fabric, and must prefer their respective positions to the alternatives of a solitary life or the joining of another group. What would be the point of staying in a group without tolerance and friendship, where one had to look continuously over one's shoulder? Hierarchies distribute not only resources but also social acceptance: the one making them arenas of competition, the other holding them together.

Not surprisingly, given this integrative function, formalized hierarchies are best developed in the most cooperative species. The harmony demonstrated to the outside world by a howling pack of wolves or a hooting and drumming community of chimpanzees is predicated on rank differentiation within. Wolves rely on each other during the hunt, and chimpanzees (at least the males, who are by far the more hierarchical sex) count on the other members for defense against hostile neighbors. The hierarchy regulates internal competition to the point of making a united front possible. The same applies to people. In a classic experiment by the social psychologist Muzafer Sherif, groups of American boys at a summer camp became more hierarchically organized and leadership oriented when pursuing a common goal, such as competition against other groups.

Hierarchies tie individuals together through a conditional proposition along the lines of "*If* you act like this, *then* we will be glad to have you." And its reverse: "*If* you do not act like this, *then* you may get punished or, worse, expelled." Thus, a low-ranking male monkey asks for trouble, so to speak, by mating with a sexually receptive female or attacking any female in the presence of the dominant male. To refrain from such behavior, and to regularly demonstrate low status, is the price he pays for undisturbed group membership. Not that subordinates lead miserable lives; it all depends on the species-typical sense of social regularity. Some species are so mild mannered and affectionate that there is no need to feel sorry for low-ranking individuals. Other species are harsher and stricter, but most of the time some measure of tolerance allows for the full integration of subordinates.[10]

How status agreement opens the door to social acceptance is shown in a nutshell during so-called *assertive reconciliations*. In the chimpanzee, for example, a high-ranking male may approach his opponent after a fight with all hair on end, locking eyes with him. If the other does not budge, another confrontation is inevitable. Most of the time, however, the other bows so as to allow the dominant to move an arm over him. The dominant will turn around immediately afterward to seal the peace with a kiss and an embrace.

Assertive reconciliations are also known in the stump-tailed macaque, in which formal rank is confirmed through a mock bite on the subordinate's wrist. The procedure is never forced on others, and subordinates sometimes approach and wave an arm under the dominant's nose to *invite* one of these inhibited bites.

Rhesus monkeys lack a similar ritual, but Spickles, an alpha male at the Wisconsin Primate Center, developed one of his own. He used it only with other males, both adults and juveniles, after having chased them. Being old and arthritic, Spickles never caught and actually punished any of these males, but as soon as they returned to the ground (they invariably escaped to the ceiling), he would walk up to them in no uncertain manner, firmly hold their head or neck, and briefly gnaw on their cheek before accepting them among the rest of the group. Sometimes the other male would turn away with a submissive grin on his face, but I have never seen one escape the I'll-tell-you-who-is-boss bite of Spickles. They endured it without protest, perhaps because Spickles had subjected them to this idiosyncratic treatment since they were little. As in the stump-tailed macaque, it was a ritualized gesture: not even the slightest injury ever resulted.

Inasmuch as assertive reconciliations require cooperation, they reflect a mutual understanding about relative ranks. Combatants come back together with just enough intimidation by the dominant to affirm his position, yet not so much that it would endanger reunification. It is the familiar link between capitulation and peace wrapped in a single package: the dominant accepts the subordinate, provided the subordinate accepts inferior status and ritual punishment.

Finally, there is the human notion of *deserving* discipline after a conflict or transgression. Not only the offended party may feel the need, but also the offender, who realizes that normalization of the relationship requires crime to be followed by punishment. Although the guilt and shame involved may be uniquely human, the assertive reconciliation of our primate relatives provided the blueprint for this process. Even between people on an equal footing, one party will make reproaches and the other will avert the eyes and apologize for any misdeeds. A temporary dominance-submission relationship is created to restore the balance.

By setting the conditions of social inclusion, the hierarchy has much in common with a moral contract. Perhaps the distinction between acceptable and unacceptable behavior represents a first step toward notions of right and wrong. An authority-based contract, though, is a rather narrow version of morality as we know it. What constitutes acceptable or unacceptable behavior within a hierarchical framework need not overlap with the distinction between right and wrong in society at large. The approval of higher-ups by no means guarantees a clean conscience: grave crimes against humanity have been committed in the name of *Befehl ist Befehl*.

It is also questionable whether, as Kohlberg thought, the earliest stages of moral development can be understood entirely in terms of obedience and conformity. Children of nursery-school age already distinguish rule violations, depending on the impact of the transgressions on others. Two developmental psychologists, Larry Nucci and Elliot Turiel, found that children consider breaking a rule that hurts others (such as stealing or lying) much more serious than a violation of mere etiquette (such as addressing a teacher by the first name or entering the bathroom intended for the opposite sex). Apparently rules that protect other people's interests weigh heavier on a child's mind than mere social conventions.

Even if respect for authority and doing as one has been told have their place in both the evolution and the development of moral capacities, they go only so far in explaining them. These tendencies need to be mixed with others, such as empathy and sympathy, before morality can emerge.

Guilt and Shame

Imagine that you are a laboratory rat in front of a dozen pieces of food. You pick up one and eat it, do the same with the next piece, and so on. When you reach for the fifth piece, a white-coated giant scares the wits out of you by clapping his hands just above your head. This nerve-racking experience is repeated trial after trial every time you touch item number five. With time, you learn to stop eating after four pieces, however tempting the remaining food may be.

Initially designed to determine the counting abilities of rats, the experiment offers a nice opportunity to look at the persistence of rules. Hank Davis describes it in an article mockingly entitled *Theoretical Note on the Moral Development of Rats*. With a slight variation on the test, he was able to prove that human-imposed rules are not held in very high regard by rats. If the experimenter is out of the room, the animals will typically pause after the fourth piece, stand on their hind legs and sniff the air, after which they happily consume all the food in sight. Davis concludes that by human standards rats are a morally bankrupt species.

The same experiment has not been tried with dogs, but there can be no doubt that these animals act differently. Dogs often obey orders in the absence of the person who trained them. Such uncoupling between behavioral suppression and the negative sanctions that instilled it is known as *internalization* of rules. While some dogs reach near-perfect internalization, even the most trustworthy dog occasionally gives in to his weaker self. After having snatched a piece of meat from the kitchen table, or having destroyed her master's shoe, the dog slinks away—ears in her neck and tail between her legs—even before anyone in the household realizes what has happened. The dog's appearance may actually draw attention to the misdeed; it betrays such awareness that we say the dog is acting "guilty."

Konrad Lorenz wrote about what he called "the animal with a conscience." The zoologist related how one of his dogs, Bully, accidentally bit Lorenz's hand when he tried to break up one of the fiercest dog fights he had ever seen. Even though Lorenz did not reprimand him and immediately tried to reassure and pet him, Bully was so upset by what he had done that he suffered a complete nervous breakdown. For days he was virtually paralyzed, and uninterested in food. He would lie on the rug breathing shallowly, an occasional deep sigh coming from his tormented soul. Anyone who had not seen the incident would have thought he had come down with a deadly disease. For weeks Bully remained extremely subdued.

Lorenz noted that his dog had never bitten a person before, so could not have relied on previous experience to decide that he had done something wrong. Perhaps he had violated a natural taboo on inflicting damage to a superior, which normally (among members of his species) could have the worst imaginable consequences. If so, instead of saying that Bully felt guilty, it might be more appropriate to say that he expected punishment, perhaps even expulsion from the pack.

Having felines and no canids at home, I never see even the slightest trace of "guilt" in my pets. Instead, I have engaged many a dog owner in debates about culpability versus anticipation of punishment, and I

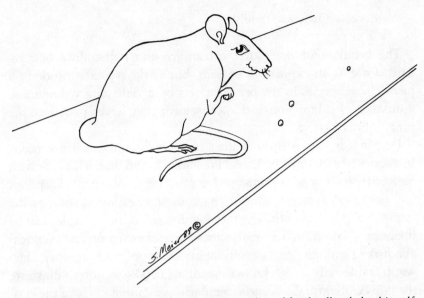

A rat trained to eat no more than a limited number of food pellets helps himself
to a contraband pellet as soon as the experimenter has left the room. Unlike humans and some other animals, these rodents do not seem to internalize rules to
any great extent. (Drawing by Susan Meier; reprinted by permission of Hank
Davis)

feel that the latter offers the better explanation. This view is supported by a test developed by an American animal behavior consult-
ant, Peter Vollmer, to convince owners of mischievous dogs that
punishment after the fact does not work.

The procedure is illustrated by the case of Mango, a Siberian husky,
who developed a habit of shredding newspapers, magazines, and
entire books. Upon returning home, the owner would take Mango to
the scene of the crime, whack her on the rump, and loudly lecture her.
Because Mango persisted in her bad behavior, acting "guilty" each
time her owner entered the house, it was believed that she knew she
was doing something wrong but continued nonetheless out of spite
over being left alone. This interpretation, however, was not corroborated by the test results.

Vollmer's test consisted simply of the owner *himself,* with Mango
out of sight, shredding a few newspapers. After having let the dog
back in, the owner left for fifteen minutes. Mango acted just as
"guilty" as when she herself had created the mess! The only thing she
seemed to understand was:

Evidence + Owner = Trouble.

The behavior of dogs after a transgression is therefore best regarded not as an expression of guilt, but as the typical attitude of a hierarchical species in the presence of a potentially angry dominant: a mixture of submission and appeasement that serves to reduce the probability of attack.[11]

Dog breeds vary greatly in their performance on obedience tasks. In a series of experiments, David Freedman found that, left alone with meat after having been punished for touching it, Shetland sheepdogs will not touch it again, whereas Basenjis start eating as soon as the trainer has left the room. The fact that internalization of rules can be bred into domesticated animals should give pause to anyone skeptical about the possibility of an evolutionary component of morality. Darwin relied heavily on his knowledge of artificial selection to illustrate the power of natural selection. Similarly, we cannot exclude the possibility that a trait subject to artificial selection in dogs has been subject to natural selection in our own species.

We are familiar with all shades of emotion related to internalization, ranging from the simple fear of getting caught to the Dostoevskian complexities of private guilt. For the moment it is unclear—and truly hard to know—if other species experience anything even close to remorse. There is no evidence that dogs ever regret forbidden acts committed away from and unbeknownst to their human masters. Not that this necessarily places them at a vast distance from us. Anticipation of punishment and fear of endangering a valued relationship are not unrelated to guilt. If rules can be internalized to the extent that they are obeyed even if the chance of punishment is minimal, fear of punishment can be internalized to the extent that we feel guilty, hence castigate *ourselves*, even if the offense will never be detected. With increased awareness of exactly what triggers the disapproval of others, and a desire to avoid such reactions even if they are not immediately forthcoming, comes the possibility of feeling culpable about transgressions regardless of who is aware of them.

What about shame? Dictionaries define it as a painful emotion caused by a sense of guilt, disgrace, or impropriety. At its core is an awareness of how one looks in the eyes of others. Could this awareness perhaps apply to Mango? After all, the mess in the house made her look bad regardless of who had caused it. Her embarrassment may seem perfectly logical to us—especially in light of her inability to explain—but it would require capacities that I am not sure we need to assume. It would mean speculation on the dog's part about how her master would take in the situation, and on whom he would blame the misdeed, given the available information. In short, it would require attribution—quite a sophisticated capacity compared to the associative learning that seems to explain her behavior equally well.

The workings of the human conscience are so intricate and rich, with their own axioms and logic, that they will forever keep philosophers, dramatists, poets, and novelists occupied. If guilt derives from the internalization of rules and values, and shame reflects concern about the opinions of others, we are dealing with complex emotions indeed. So complex, in fact, that the term "emotion" does not do them justice; self-consciousness, perspective-taking, and attribution are also involved. Even the so-called emotivists, who tried to derive morality from sentiment, recognized the need for this cognitive component. Adam Smith imagined that our every thought and deed is monitored by an "impartial spectator," as if we are holding up a mirror to decide how our behavior looks from the outside. Thus, the social environment that shapes conscience in childhood is gradually replaced by a set of internal regulators that mimic its effects. We need wait no longer for the praise and blame of others: we mete them out to ourselves, according to the standards we have made our own.

In my more cynical moments I wonder if we do not overrate the power of internalization. Consider how people throw inhibitions overboard when conditions change, as during war or famine, during political unrest, or when running with a rowdy mob. Many a supposedly upstanding citizen loots, steals, and kills without compunction when there is little chance of getting caught or when resources have become scarce. Even a less dramatic change of circumstances, such as a vacation in a foreign land, may induce people to act in silly or outrageous ways unthinkable in their hometown. Are the members of our species—apart from the occasional saint—really any better than the rats who sniff the air and decide that it is safe to do what once was forbidden? Perhaps guilt and shame are less deeply rooted than we like to think; fear of negative consequences may need to persist, at least at the back of our heads, for our moral reasoning to stay on track. It might be argued, then, that the link between external and internal regulators of behavior is never completely lost.

One of the best-known external regulators in social primates is the effect of high-ranking males on the sex life of low-ranking males. The relation between punishment and behavioral control discussed for dogs, including signs of "guilt," can be observed. As a student working with long-tailed macaques, I followed activities in an outdoor section of their cage connected to the indoor section by a tunnel. Often the alpha male would sit in the tunnel so as to control both sides. As soon as he moved indoors, other males would approach the females outdoors. Normally they would be in deep trouble for doing so, but now they could mate undisturbed. Fear of punishment did not disappear: I have seen low-ranking males act extremely submissive, with wide grins on their faces, if they encountered the alpha male shortly after one of these sneaky copulations—even though alpha could not possibly have known what had happened.[12]

The effect of past behavior on current relations was put to the test by Christopher Coe and Leonard Rosenblum. In one series of experiments, subordinate male macaques were allowed to be with a female while the dominant male looked on from inside a transparent isolation chamber. None of the subordinate males dared to initiate sex under these circumstances. Their attitude changed dramatically, however, in the second test series, in which dominant males were entirely removed. Now the same subordinates felt free to copulate. They also suddenly performed bouncing displays and walked around with their tails proudly in the air, actions characteristic of high-ranking males.

Upon return of the alpha, these subordinate males showed considerably more avoidance and submission than when they had not been with a female. The investigators conclude that "this experiment provides intriguing evidence that animals can incorporate behavioral rules which are associated with their social role and can respond in a manner that acknowledges a perceived violation of the social code."[13]

These kinds of when-the-cat-is-away situations are amusing to watch, as the cat is never far from the minds of the mice. In the rhesus group at the Wisconsin Primate Center, Spickles sometimes seemed to get tired of keeping an eye on five or six restless males during the breeding season. Or perhaps he wanted to warm his old bones in the heated indoor section. At any rate, he would go inside once in a while, sometimes for half an hour at a stretch, leaving the others plenty of opportunities to mate. The beta male, Hulk, was popular with the females and often mated on these occasions. He was so obsessed with Spickles' whereabouts, however, that he was irresistibly drawn to peek inside through a crack in the door. Perhaps Hulk had had unfortunate experiences in which the boss unexpectedly showed up, and he wanted to make sure that alpha was staying put. Inasmuch as rhesus monkeys are multiple-mount ejaculators (several mounts and dismounts are required before the male ejaculates), Hulk would nervously race back and forth a dozen times between his partner and the door before completing a mount series.

These observations suggest that social rules among primates are not simply obeyed in the presence of dominants and forgotten in their absence. If this were true, Hulk would not be checking on Spickles, and low-ranking males would not be overly submissive after their exploits. The responses show that inhibitions are rooted deeply enough that concern about the enforcer's reaction persists in his absence. Admittedly, this is not much of an internalization compared to the near-total behavioral control that can be instilled into animals artificially bred for obedience. It may nevertheless have provided the starting point in the primate lineage for the evolution of a capacity for guilt and shame.

Unruly Youngsters

How monkeys acquire social rules is largely unknown; study of this topic may produce fascinating parallels to human moral development. Initially, comparisons would of course suffer from a problem pointed out by Davis in his paper on rats, namely that descriptions of human morality tend to be "couched in terms that have limited operational relevance to other species."[14] Perhaps animal research will stimulate psychologists to come up with definitions that do not rely exclusively on verbalized thoughts and feelings but also cover how people behave. What is different about the way we *act* that makes us, and not any other species, moral beings? And if this question is hard to answer, what exactly is the significance of morality? Nothing like a little challenge to bring matters into focus! Similarly, language research on apes has served to sharpen the definition of language, if only because of the desire of linguists to keep their domain free of hairy creatures.

John Finley Scott, a sociologist, sides with biologists by putting the need for a new conceptualization this way in *Internalization of Norms*:

A definition of the norm that applies to all learned social organization, one that can build on interspecific comparisons, will serve sociological theory better than an anthropomorphic concept which imposes purely terminological barriers across an otherwise continuous dimension of activity.[15]

Monkeys do not come into this world with social rules imprinted in their heads. Neonates enjoy a certain *Narrenfreiheit*: like jesters at the court, they live above the law. Young infants may bump into high-ranking adults or approach food that others are interested in without being threatened or chased away as any juvenile would. In our studies of rhesus monkeys, for example, we discovered that infants occupy a better position than their mothers in the group's drinking order because they are allowed to approach the water basin and drink side by side with dominants who do not tolerate the mothers. Adult males accept them better than do high-ranking females, and the youngsters quickly learn to approach the former and not the latter.

Mothers seem aware of these special privileges. Fernando Colmenares observed a female baboon at the Madrid Zoo who held her infant firmly by the tail at feeding time while the infant entered a dominant male's personal space, normally taboo when food is around. The infant was not terribly interested in the food, but played with pieces picked up near the male. The mother, in turn, pulled her offspring toward herself a few times to quickly remove these tidbits for her own consumption. Noticing this, the male threatened her (not the infant), at which the mother presented her behind and kept the food. The infant served as a tool to circumvent the rules.

Infants do not always respond to threats, and perhaps as a result adults exaggerate the signals. Instead of slightly raising the eyebrows and giving them a stare—usually sufficient in the case of a juvenile—they will perform the entire pattern very clearly, with staring eyes, open mouth, ears spread, head bobs, and so on. I have seen adults grasp an infant's head with both hands to threaten it at close range straight into the face. They may have learned to give such "instructional" threats because very young infants often blithely ignore less conspicuous warnings.

With age, the consequences of ignoring threats become more severe. Punishments move from a slap or a firm shake with the hand to mild bites, until one day the infant receives its first serious bite. The time it can get away with murder is forever gone: the young monkey learns which situations and individuals to avoid. The risk of attack differs by species and is perhaps greatest in the rhesus monkey, where infants occasionally lose fingers and toes as a result of punishment. With the cost being so high at such a young age, it is no wonder that this species heeds its hierarchy so carefully.

Like Spickles, Orange, the alpha female of the Wisconsin rhesus group, developed her own stereotypical punishment. The big difference was that whereas Spickles' gnaw on the cheek was inhibited, Orange's was a real bite. Curiously, she used it almost exclusively on infants approximately six months old. At an unexpected moment, Orange would grab them and clamp her teeth so firmly on their wrist that she would draw blood. The victim would be upset afterward, limp for a couple of days, and from then on keep a safe distance and pay close attention to Orange's movements. What made the deepest impression was perhaps not even the pain itself, but the fact that no one had come to help. As all-powerful queen, Orange could do whatever she wanted. During the decade that I followed the group, this became a highly predictable pattern; few infants escaped the treatment. Thus, each fall (rhesus monkeys are seasonal breeders) we would see Orange do her wrist-biting (or notice its results) as if she had determined that it was time to instill fear and order into the new generation.

There is no single individual from whom infants and juveniles receive more aggression, however, than their own mother. Usually, of course, it is of the nondamaging kind, but bites and even injuries do occur. Irwin Bernstein, a well-known American primatologist, interprets it as *socialization,* in which mothers teach their offspring to inhibit particular behaviors that may get them into trouble. Even though maternal aggression may not be to the youngster's immediate advantage, it promotes the caution and behavioral control required for survival in a hierarchically structured social environment.

As with the implementation of rules, it is questionable that this socialization process is intentional. Undoubtedly, the young monkey learns lessons for life from the corrections received, but adults are not necessarily teaching them these lessons on purpose. As yet the issue is unresolved and relates to how well animals understand the effects of their own behavior. Do they realize that a punitive action may permanently alter the other's behavior? Do they evaluate the results and punish harder if previous attempts failed? And why do they sometimes give up and let things be, as we saw with some of the handicapped monkeys? Azalea, for example, never achieved a well-defined rank. Her responses lacked consistency; she often wavered between threat and retreat. Yet the others simply ignored this behavior. Instead of punishing Azalea for threatening back at them, dominant monkeys would stare at her, perhaps make a brief lunge, then leave her alone as if she was just not worth wasting energy on.

We also should not assume that social rules are always actively enforced by one party and passively learned by another. The establishment of rules is a dynamic process involving rule-testing and social exploration on the part of the learner.

One technique is *teasing*. Because of the different risks involved if a teaser is caught, this behavior is rare in rhesus, not uncommon in stump-tails and baboons, and ubiquitous in chimpanzees. Youngsters of the last species throw handfuls of dirt or pebbles at their elders, hit them with sticks, splash them with water, jump on their heads when they are dozing, and so on. Much of the time, the individual thus bothered takes it remarkably well, tickles the youngster, or makes a mock chase that turns the whole incident into a game. The individuals who cannot resist a hostile reaction are the ones who are going to be pestered more, as Otto Adang discovered in his study of teasing by young chimpanzees at the Arnhem Zoo.

Adang documented the various teasing techniques, the reactions to them, and how they change with age. At one end of the spectrum are the innocuous little pushes from behind by an infant, who jumps away when the adult turns around. But at the other end are full-blown charging displays of adolescent males seeking to engage females in physical combat. Throwing sand into the faces of those in power is a sure way of provoking a reaction—unpleasant stimuli have this advantage—and learning about one's place in society. Where do I stand, what are my limits? Provocation is also a way of expanding limits by monitoring carefully how others react, and looking for an opening to cow them.

In short, teasing serves to gather information about the social environment, and to investigate authority. Its presence in primate youngsters confirms the importance of knowing the limits set by society.

The Blushing Primate

Even if rule-learning, rule-testing, internalization, and guiltlike behavior are observable in our closest relatives, it is evident that no species has perfected these processes to the degree that we humans have. Whereas some other animals can be described as conformists—that is, they learn to obey rules—people go well beyond conforming by internalizing not merely the rules but also the values and ideals behind them. In addition, we build such an elaborate network of sanctions that even when the coast seems perfectly clear, transgressions entail serious risks, ranging from gossip by an accidental witness to entanglement in lies. Futhermore, millions of people believe in the watchful eye of an omnipresent God, in retribution after death, and in a connection between moral decay and natural catastrophes. There is no escape: we are subject to both real and imagined pressures to behave in specific ways.

Our extreme susceptibility to social and religious influences is closely tied to the central role of *reputation* in human society. Reputation yields both the stick and the carrot for much of our behavior. Evolutionary biologists, such as Richard Alexander, relate this circumstance to systems of mutual aid that are constructed around trust. Within such systems, partners are selected on the basis of their ability to make commitments. Inasmuch as every act is indicative of future acts by the same person, it pays to watch others closely and learn what to expect from them. This scrutiny in turn produces a definite preoccupation with the social mirror, to the point of causing constant worry about possible loss of face.

If personal reputations varied from day to day, there would be little reason to adhere to social norms. Why not pay lip service to the rules in public but break them whenever convenient? Yet the system does not work this way. Reputations are durable and incredibly fragile: built up over many years, they can be destroyed by a single faux pas. The television evangelist who slept with a secretary is history, as is the political candidate who cheated on his taxes. And imagine the situation of our ancestors, who undoubtedly belonged to the same small community for most of their lives: virtually everything there was to know about them must have been known by virtually everyone.

In order to become a respected member of the community, we need to be almost blindly consistent. To resist every little temptation to stray from the chosen path, we need to believe steadfastly in right and wrong. Only a firm moral conviction ensures behavior that contributes to a lasting reputation of honesty. Indeed, most people reach a point at which the values imbued by society combined with their personal experiences crystallize into a stable pattern of thought and behavior from which they cannot deviate without great personal discomfort. Instead of being led by the reactions of others, or responding to immediate situations, to stay on course they rely on an internal compass, bolstered by strong emotions of guilt and shame.

It is this inner strength that we look for when evaluating a person's trustworthiness; we are experts at detecting the difference between people with an effective compass and those with a defective one.

According to Robert Frank, an American economist who has written extensively about reputation and emotional commitment, it hardly pays to fake generosity or obey rules only in public. People are too astute at distinguishing such a facade from acts that come from deep inside. In *Passions within Reason* Frank explains that people committed to integrity and fairness occasionally pass up opportunities to achieve private gain, yet create other opportunities for themselves that are out of reach of opportunists. For example, an honest man will leave a tip in a restaurant that he does not expect to visit again, whereas the opportunist will see this as a chance to save money. In the short run the honest man loses, but in the long run the convictions dictating his behavior will be recognized and valued, providing him with a passport to partnerships with like-minded people.

And there is more: reputation-building has apparently been so central in our evolutionary past that serious obstacles have been erected for anyone trying to take the unethical route. Painful truths sometimes emerge because of our incomplete control over voice, eyes, and the small arteries of face and neck—exactly the body region where a crimson color best signals guilt and shame.

Blushing is a truly remarkable trait, as Darwin long ago realized: "Blushing is the most peculiar and the most human of all expressions. Monkeys redden from passion, but it would require an overwhelming amount of evidence to make us believe that any animal could blush."[16]

What could be the advantage of signaling shame or embarrassment? Does blushing not seriously handicap us? Quite a few additional assumptions need to be made before this characteristic fits the self-interest models of evolutionary biology, assumptions that include the need to come across as a desirable partner for cooperative ventures—one who can be relied on, thanks to a strong internal compass that makes that person susceptible to guilt and shame. A person who lies without blushing, who never shows remorse, and who grabs every opportunity to bypass the rules just does not strike us as the most appealing friend or colleague. The uniquely human capacity to turn red in the face suggests that at some point in time our ancestors began to gain more from advertising trustworthiness than from fostering opportunism. And what more effective way to do so than by telltale signs beyond their control?[17]

In our analysis we should not confuse evolutionary benefits with actual motives. Because we are unable to change color on command, blushing cannot possibly be part of a calculated strategy to create a favorable impression. People assimilate the values of their society so profoundly and so completely that they are genuinely moved by them. I firmly believe that people can be altruistic and honest without on every occasion thinking of the advantages.

This view of human motives as independent of the evolutionary process that shaped our minds and bodies is, as we have seen, not universal among biologists. Some believe that a self-serving agenda guides human behavior under all circumstances; people who believe otherwise are simply fooling themselves. To this day the sincerity of human feelings continues to be a topic of debate. As recently as 1994 Robert Wright, an American science writer, argued in *The Moral Animal* that humans are potentially but not naturally moral. In Wright's eyes, we are hypocrites living in constant denial of our thoroughly selfish nature.[18]

To give the human conscience a comfortable place within Darwin's theory without reducing human feelings and motives to a complete travesty is one of the greatest challenges to biology today. We are born to absorb rules and values, many of which place community interests above private ones. We have been selected for such profound internalization that these rules and values become literally our own. We have built-in physiological mechanisms that frustrate attempts to cheat. No doubt these capacities evolved because they served a purpose in the highly cooperative and trust-based societies of our ancestors. To believe that this purpose must figure as a conscious or unconscious motive in our minds assumes a direct connection between genetic adaptation and day-to-day decision-making for which there is absolutely no evidence. The same assumption would lead one to claim that if squirrels store nuts in the fall they must know about the deprivations of winter and spring. Squirrels would have gone extinct long ago if such knowledge were required, and people would never have developed a conscience had their minds been preoccupied with the reproductive calculations that fascinate evolutionary biologists.

Two Genders, Two Moralities?

Like every European boy, I could barely walk when I began to play the world's most popular sport. Before I knew it, along with the fun of chasing the leather, I was receiving important moral lessons. Team sports, with their rules and expectations, are a microcosm of society. Whether one plays on grass or on pavement, the risk of a fall is not nearly as serious as the risk of not knowing how to behave, or not contributing to the team. Most important, it is not the almighty adults who keep an eye on you and on whom you keep an eye; it is your peers and equals!

In the beginning are simple rules about which body parts may touch the ball and what kind of contact with other players is within bounds. Gradually the rules become more complex and precise, until one day an older boy shouts "offside" during a promising offensive—and the most frustrating regulation of all is introduced. In addition to obeying these rules, you learn to uphold expectations and obligations. On the one hand, you are not meant to simply stand around and take it easy. On the other hand, you should not try to outshine your pals. Instead of shooting at the goal from an impossible angle, for instance, you are supposed to pass the ball to a teammate who may end up triumphant. He, in turn, is expected to acknowledge your assistance.

This knowledge is acquired via endless debates about what one did versus what one should have done. And even if the game is not halted for a particular violation, it is made perfectly clear that the disadvantaged team has been generous—a gesture to be recalled as soon as the other team fusses over some minor infraction.

Boys seem to enjoy these legal battles every bit as much as the game itself, as observed in 1972 by Janet Lever in a now-classic study of children's games in Connecticut. Lever found that girls tend to play in smaller groups, and less competitively, than boys. Their games are considerably shorter, partly because girls are not as good as boys at resolving disputes. Based on observations and interviews, Lever contrasted the two sets of attitudes toward disagreement: "Boys were seen quarreling all the time, but not once was a game terminated because of a quarrel, and no game was interrupted for more than seven minutes." In contrast "Most girls claimed that when a quarrel begins, the game breaks up, and little effort is made to resolve the problem."[19]

Following in the footsteps of the famous Swiss developmental psychologist Jean Piaget, who first analyzed the moral lessons derived from rule-bounded games, Lever concluded that boys' games offer better preparation than girls' games for the adjudication of disputes, respect for rules, leadership, and the pursuit of collective goals. With jump rope and hopscotch involving turn-taking rather than contest, and with girls playing in pairs or trios of close friends rather than in larger groups, the typical girls' games seem to serve as a training ground for delicate socioemotional skills. Lever saw these skills as valuable mainly for future dating and marriage, not as part of moral development.

Certainly Carol Gilligan has another interpretation. In *In a Different Voice* the American psychologist claims that female moral commitment is rooted in attachment, intimacy, and responsibility for the other, whereas male commitment is oriented to rights, rules, and authority. For simplicity, let us call these types sympathy-based morality and rule-based morality. Gilligan makes the point that although human morality is based on both rules and sympathy, men and women reach an integration of the two via different routes. For this reason, the sensitivity skills developed in girls' games may be as morally relevant as the boys' experience with conflict resolution and fair play.

Instead of standing in awe before principles higher than themselves, women have a rather down-to-earth approach to the dilemmas of moral judgment presented by researchers. Gilligan asked her subjects how they would resolve an imaginary situation in which the interests of various persons collided. Female subjects wanted to know all sorts of missing details concerning the nature of these hypothetical people—where they lived, how they were interconnected, and so on. They ended up with moral solutions tailored to these people's needs based on the logic of social relationships rather than abstract principles.

Gilligan warns against the excesses of a rule-based morality: "The blind willingness to sacrifice people to truth has always been the danger of an ethics abstracted from life."[20] In a powerful biblical reference, she contrasts Abraham's offering of his son's life to show the strength of his faith with the woman before Solomon who lied about her motherhood in order to save her child. In his wisdom, Solomon knew that only a mother would do such a thing.

In one study on this topic, Kay Johnston asked adolescents to suggest solutions to Aesop's fables, such as the one in which a porcupine, on a cold night, enters a cave occupied by moles. "Would you mind if I shared your home for the winter?" asks the porcupine. The moles consent, but soon come to regret their generosity; the cave is tiny, and the porcupine's quills scratch them at every turn. When they finally find the courage to ask their visitor to leave, the porcupine objects, saying "Oh, no. This place suits me very well."

A majority of the boys who read this fable preferred solutions based on rights ("The porcupine has to go; it's the mole's house" or "Send the porcupine out, since he was the last one there"), whereas most girls preferred mutually agreeable solutions ("Wrap the porcupine in a towel" or "The both of them should try to get together and make the hole bigger"). A fair number of children proposed both kinds of solutions, however, and many were able to switch orientation when asked if there was another way of solving the problem. Even if the two sexes set different priorities, both were capable of following alternative lines of reasoning. This seems to be true to such a degree that many of Gilligan's colleagues have begun to question her dichotomy between male and female moral orientations.[21]

For one thing, it would be quite incorrect to say that men attach little value to expressions of sympathy. Even if men typically favor a rights orientation, we should not forget that rights are often an abstract, regulated form of caring. Rights go hand in hand with duties, hence are concerned with social relationships and responsibilities as well. It is to male philosophers, such as David Hume, Arthur Schopenhauer, and Adam Smith, that sympathy owes its prominent place in moral theory. Conversely, it cannot be said that women are aversive to framing issues in terms of rights, as the last few decades of worldwide stuggle for equity, justice, and fairness ("women's rights") have shown. In other words, the two orientations distinguished by Gilligan are not quite as far apart as they sound.

The centerpiece of the attack on her position, however, is a comprehensive review of research on moral development in more than ten thousand subjects by Lawrence Walker. Failing to find systematic evidence for sex differences in the data collected by a large number of investigators, Walker concludes that "the moral reasoning of males and females is more similar than different."[22]

Walker goes on to admit, however, that the moral development scales used for his analyses say very little about the precise norms and social orientations underlying moral judgment. Could it be, then, that both sides in the debate have a point? The intuitive appeal of Gilligan's distinction between male and female attitudes should give pause to her critics. When an American linguist, Deborah Tannen, in *You Just Don't Understand* classified the conversational styles of men and women in terms of a male preoccupation with autonomy and status and a female preoccupation with intimacy, she touched on very much the same issues. Tannen's position, too, struck a chord. Critics argue that the proposed sex differences represent convenient "stereotypes" that people cannot resist, but it is of course also possible that these ideas have become popular because they reflect some deeper truth about ourselves.

At least one sex difference manifests itself so early—on the very first day of life—that culture can be ruled out as an explanation. Newborn babies cry in response to the sound of another baby's cry. It is not just a matter of noise sensitivity, as babies react more intensely to these sounds than to equally loud computer-simulated cries or animal calls. Their reaction is regarded as an expression of emotional contagion, and thought to provide the underpinnings of empathy. Studies consistently demonstrate higher levels of this kind of emotional contagion in female than in male infants.

At later ages as well, empathy seems more developed in females. On the basis of a wide range of studies, Martin Hoffman concludes that whereas the sexes are equally capable of assessing someone else's feelings, girls and women are more strongly *affected* by the resulting knowledge: "Females may be more apt to imagine that what is happening to the other is happening to them; or, more specifically, to imagine how it would feel if the stimuli impinging on the other were impinging on the self."[23]

It is possible that boys and girls play different games, build different social networks, and develop different moral outlooks because they are born with psychological differences that are subsequently elaborated and modified by the social environment. In this view, culture and education mold gender roles by acting on genetic predispositions. If true, two fundamental questions result. First, Why do people seem to encourage these particular sex differences? Why are girls taught to be sensitive and accommodating, and boys urged to go out and prove themselves in competition with other boys? Is it easier to amplify what is there than to steer social attitudes into a new direction, or are all human societies structured in such a way that it is convenient or necessary to have men and women occupy these particular roles? This question is one for developmental psychologists, sociologists, and anthropologists.[24]

The second question concerns the evolutionary reason for these differences. Why would females be born with a stronger responsiveness to someone else's feelings, and why would males have an orientation toward competition and status? Do such differences make biological sense? The short answer is yes. The long answer is that it is always dangerous to think in typologies, and that we are basically comparing *average* tendencies. Yet these tendencies over the long run may have favored the reproductive success of males, who depend on status to gain access to mates, as well as females, whose offspring would stand no chance at survival without continuous care and attention.

By bringing the issue to this level, we are no longer comparing moral outlooks, but instead basic psychological traits that may have served males and females in the course of evolution.

Umbilical versus Confrontational Bonds

Even if contemporary society emphasizes the considerable nurturing abilities of men, it cannot be denied that in virtually all primate species females are the primary or only caretakers of the young. Responsibility for a completely helpless and vulnerable member of the species, who depends on you for food, transport, warmth, comfort, and protection, is an obvious starting point for the evolution of sensitivity to the needs of others. The slightest distress call, often barely audible, alerts the mother and causes her to adjust either the infant's position or her own posture so as to allow access to her nipples, stabilize the infant on her belly or back, or remove pressure. Failing to do so may cost the infant's life, as we sadly learned from a deaf chimpanzee who never managed to raise offspring. Despite her fascination with the infants, she would accidentally sit on them without noticing, or deprive them of milk for extended periods without paying attention to their complaints.

Attached with an emotional umbilical cord to her offspring, the primate mother is never free. Her infant may be asleep or nursing when she is ready to move, or she herself may be resting when an infant-crazed, clumsy juvenile female tries to kidnap her little one. Then there is the sheer weight when the offspring grow bigger. It may be fetching to see an infant monkey hold on with hands and feet to its mother's hair, hanging upside down beneath her, yet that infant must be a painful load. Often females risk trouble with their offspring, trying to pry open the clenched little fists; the infant may protest noisily as it has no desire to give up the free ride. Torn between the need to move and the impossibility of leaving their offspring behind, mothers almost always give in and carry the burden.

Even when the young begin to move on their own, mothers need to stay alert in order to protect them against aggression or predation, and to help out when their motor skills fall short. In some arboreal species mothers use their own bodies as bridges, holding on with their feet (or tail) to the branches of one tree and with their hands to those of another. I once saw a muriqui monkey do this high in the canopy of a Brazilian forest. American primatologist Karen Strier has seen many such live bridges, describing them as follows:

> Juveniles screech when their mothers leave them to negotiate difficult tree-crossings alone. Small juveniles are reluctant to fling themselves across wide gaps in the canopy which larger muriquis can swing or leap across with ease. Often in these contexts, mothers respond to their offspring's screeches by using their bodies to form bridges for them to cross, and, in the early stages of weaning, young muriquis take advantage of these opportunities to latch onto their mothers. Later on, however, they run across their mothers' backs without even trying to hitch a ride.[25]

Given that nurturance is essential to raising offspring, and that in nearly two hundred million years of mammalian evolution the need to provide it has applied to each and every mother—from the tiniest mouse to the largest whale—it is no mystery why females of our own species value intimacy, care, and interpersonal commitment. These traits are visible in parenting styles (with maternal love being unconditional and paternal love being more qualified), but also, as Gilligan points out, in the approach to moral issues. Impersonal rights and wrongs are not a top priority of women, who often favor compromises that leave social connections intact.

The moral outlook of men, based on rules and authority, follows directly from a dominance orientation. Wherever men come together—in the military, in secret societies, in religious organizations, in prison, in corporations—hierarchical relations are rapidly established to create an environment in which men seem to work together best. Despite its grounding in competition, the hierarchy is essentially a tool of cooperation and social integration. Picking a fight can actually be a way for men to relate to one another, check each other out, and take a first step toward friendship. This bonding function is alien to most women, who see confrontation as causing rifts. If an open fight does break out, boys and men are more likely to make up afterward, as illustrated by Janet Lever's playground study. Similarly, a Finnish research team found that grudges among girls outlast those among boys,[26] and Tannen reports hostile conversations followed by friendly chats among men:

> To most women, conflict is a threat to connection, to be avoided at all costs. Disputes are preferably settled without direct confrontation. But to many men, conflict is the necessary means by which status is negotiated, so it is to be accepted and may even be sought, embraced, and enjoyed.[27]

In chimpanzees, too, males are the more hierarchical sex and reconcile more readily than females. Females are relatively peaceful, yet if they do engage in open conflict the chances of subsequent repair are low. Unlike males, females avoid confrontation with individuals with whom they enjoy close ties, such as offspring and best friends, whereas they let aggression run its destructive course in case of a fight with a rival. During group formation in captivity, females have been observed to reconcile at higher rates—making peace is well within their abilities—but typically, in well-established groups, it is the males who go through frequent cycles of conflict and reconciliation, who test and confirm their hierarchy while at the same time preserving the unity required against neighboring communities.[28]

Thus, chimpanzees and humans seem to share fundamental sex differences in their orientation toward competition, status, and the preservation of social ties. In both species, however, simple dichotomies should be treated with caution as there is a great deal of plasticity. With the exception of nursing, males are capable of every behavior typical of females, and vice versa. In chimpanzees, for example, males have been known to adopt and care for orphaned juveniles, and females have been known to intimidate others by means of charging displays as impressive as those of males. It all depends on the circumstances. Most of the time, sex differences follow a specific recognizable pattern; yet in an environment that requires different responses, both sexes can and will adjust.[29]

Similarly, in modern society we see households run by single fathers and we see women working their way to the top. It is this culture dependency of sex differences that has lured feminist scholars and social scientists—at least for a time—into thinking that biology is irrelevant, as if the manifest influence of one factor would in any way preclude the influence of another.

Mindful of these complexities, Gilligan softens her position toward the end of her book, arguing that with increasing maturity both sexes move away from the extremes. Women move from the absolute of care, initially defined as not hurting others, to the inclusion of principles of equality and individual rights. Men begin to realize that there are no absolute truths, and that not all people have equal needs. The result is a more qualified judgment and an ethic of generosity. The intertwining of these strands of morality into a single basket is partly a product of men and women learning from each other that there are different angles from which problems can be tackled, and that morality can be reduced neither to a book of rules nor to pure warmth and sympathy.

Primus inter Pares

If men are such hierarchical creatures, how do we account for what anthropologists call "egalitarian" societies, such as the Navajo Indians, Hottentots, Mbuti pygmies, !Kung San, and Eskimos? Ranging from hunter-gatherers to horticulturalists, many small-scale societies are said to eliminate distinctions—other than those between the sexes, and between parent and child—of wealth, power, and status. The emphasis is on equality and sharing. Since it is entirely possible that our ancestors lived in this fashion for millions of years, could it be that hierarchical relations are less prototypical than assumed?

Status differences are never completely abandoned, however. Instead of having no hierarchy at all, egalitarian societies occupy one end on a spectrum of dominance styles ranging from tolerant to despotic. *Dominance style* refers to the amount of control that high-ranking individuals exercise over low-ranking ones—and vice versa. An egalitarian dominance style is produced by political leverage from below that restrains the top's power and privileges. If our species' sense of social regularity goes by the name of a sense of justice, it is precisely because we possess this equalizing tendency. We sometimes get rid of unpopular power holders, or at least criticize them and let them understand that there are limits to our obedience. Admittedly, there has been little evidence for egality during long periods of history—equity and justice are sometimes thought of as recent innovations—yet the existence of so-called egalitarian societies suggests that the tendency has been with us for quite some time.

Variation in dominance style is obvious even within such a relatively small cultural sample as Western Europe. The inhabitants of some regions have a reputation for being almost militaristically hierarchical, others for maintaining sharp class divisions, and yet others for cutting down to size anyone inclined to self-aggrandizement. There may be ecological explanations. Odd as it may sound, the dislike of the Dutch for aristocracy has been linked to life below sea level. The continuous struggle to keep the country dry created strong common purpose, particularly during the onslaught of storm floods in the fifteenth and sixteenth centuries. Every burgher had to pitch in and carry bags of clay in the middle of the night when a dike was near collapse. Those who put status above duty were in effect disqualified from the benefits. The contempt voiced by a well-known engineer of that period for "the slippers, the tabards, and the fine fur mantles which have no value at the dikes"[30] became so much part of Dutch culture that even today its monarchy shuns pomp and circumstance.[31]

Paralleling this cultural variation, differences in dominance style are recognizable within the primate order. Once again, it was Abraham Maslow who first drew attention to the phenomenon. Publication of his observations was delayed for years because prominent primatologists of his time disagreed with his views. (Maslow alludes to their objections without revealing their exact nature.) Maslow was entirely correct, though, to contrast what he termed the "dominance quality" of chimpanzees and rhesus monkeys. Two more different temperaments can scarcely be found.

In rhesus, Maslow explained, the dominant puts priority rights above all else and is never the least bit shy about punishing a subordinate for liberties taken. The subordinate, in turn, lives in fear. What a difference with the chimpanzee, in which the dominant often acts as friend and protector of the subordinate: they share their food, and a frightened subordinate may run straight into the arms of the dominant. There is also remarkable room for low-status individuals to voice their feelings. With some delight Maslow recounts how a young chimpanzee got so fed up with attempts of the most dominant ape to take her friend away to play that she ended up attacking the big male and hammering him with her fists. The only reaction he showed was to run away laughing (chimpanzees make hoarse laughing sounds during play) as if her audacity amused him! Maslow rightly notes that the same protest by a subordinate rhesus would have cost her dearly.

Maslow's study had serious limitations, however, not least because his chimpanzees were rather young (which may be what the referees of his paper objected to). Yet even the rank order of adult chimpanzees is never as strict and rigid as that of rhesus monkeys. Real and profound differences are glossed over by flat statements that all primates know dominance-subordination relationships.

To appreciate this variability we do not need to compare such distant relatives as rhesus and chimpanzees. Rhesus monkeys are despotic even when compared to some members of their own genus, such as stump-tailed macaques. Stump-tails seem to emphasize harmony over priority rights. Compared to his rhesus counterpart, a dominant stump-tail tolerates quite a bit more, including counterthreats in an argument, and drinking or feeding from the same source shoulder to shoulder with subordinates. Stump-tails also groom one another more and are among the best peacemakers after fights.

According to socioecologists, who look at the natural environment to explain social dispositions, the optimal condition for the evolution of egalitarianism is dependency on cooperation combined with the option to leave the group. The first prerequisite is not unlike the struggle of the Dutch against the ocean: there is nothing like a common enemy to promote unity and tolerance. The need for cooperation may relate to the physical environment, to predators, or to enemy groups of the own species. The second stipulation—freedom to exit—depends on predation risks outside the group and the willingness of neighboring groups to accept new members. If there exist both mutual dependency and realistic opportunities to leave, dominants had better be very "nice" to their companions lest they find themselves without anyone to dominate. The reverse is, of course, equally true: when subordinates have nowhere to go, dominants can exploit and terrorize them with impunity.[32]

Unless, that is, subordinates have a way of joining forces: alliance formation, a characteristic of the primate order, blunts absolute power. Originally, alliances chiefly served the acquisition of dominance, as when female monkeys assisted their younger kin to ascend the social ladder. The same instrument, however, permits lower echelons to rise up against and even overthrow rulers. Such use is already visible in the chimpanzee, in whom cooperation against dominants is not nearly so unusual as in macaques and baboons. Thus, alliances evolved from a means of building and enforcing the hierarchy, to one capable of weakening it.

In egalitarian societies, men intent on commanding others are systematically thwarted in their attempts. The weapons used by their supposed inferiors are ridicule, manipulation of public opinion, and disobedience. The boastful hunter is cut down by jokes about his miserable catch, and the would-be chief who tries to order others around is openly told how amusing his pretensions are. The power of leaders is thus delineated by an alliance from below. Christopher Boehm, an American cultural anthropologist (who later took up research on chimpanzees), studied these leveling mechanisms. He found that leaders who become too proud or bossy, fail to redistribute foods and goods, or close their own deals with outsiders, quickly lose respect and support.[33]

In extreme cases they pay with their lives. A Buraya chief who appropriates the livestock of other men and forces their wives into sexual relations risks being killed, as does a Kaupaku leader who oversteps his prerogatives. Such executions, sometimes secretly arranged, have been reported for ten of the forty-eight cultures reviewed by Boehm. The author goes so far as to speak of a "reversed hierarchy," meaning that leaders are in effect dominated by their followers. Other scientists have objected to this notion on the grounds that it ignores the fact that, however much the hierarchy is leveled, a leader is a leader nonetheless. David Erdal and Andrew Whiten explain: "Since respect is still given to leaders in particular situations, the incipient hierarchy is not really reversed but rather prevented from developing beyond those particular situations where leadership is required."[34]

In other words, egalitarian societies permit certain men to act as leaders because it is harder to survive without any leadership at all than with a limited degree of inequality. One domain in which the need for leadership is most strongly felt is the resolution of intragroup disputes. Instead of having all community members take sides—and make matters worse—what better way to handle the situation than by investing authority in a single individual assigned to enforce the interests of the community by finding fair solutions to disagreements?

Conflict mediation from above is not limited to our species. Irwin Bernstein was the first to describe the *control role* of monkeys: alpha males respond aggressively to external disturbances, such as predators or other groups, but also internal disturbances, such as conflicts within the group. Smothering the latter is achieved through direct interference, often by chasing off the aggressor. There is great variation in the effectiveness of control measures, however. I have known alpha males who could stop a fight by raising an eyebrow or with a single step forward, whereas others only aggravated the situation by getting involved. Competent control requires special skills, such as interceding with just the right amount of force—enough to command respect, yet not so much that everyone would have been better off left alone.

Individual variation in the effectiveness of intervention, combined with the fact that the control role is poorly developed in some commonly studied species, has led to serious reservations about the whole concept. All the same, a few species do exist (stump-tailed monkeys, gorillas, and chimpanzees) for which there can be little doubt that high-ranking individuals consistently and effectively mediate conflicts. The strongest evidence concerns the chimpanzee, whose pacifying interventions have been documented at Gombe National Park by Boehm and in captivity by me. Dominant chimpanzees generally break up fights either by supporting the underdog or through impartial intervention. All hair on end, they may move between two combatants until they stop screaming, scatter them with a charging display, or literally pry locked fighters apart with both hands. Their main objective seems to be to put an end to the hostilities rather than to support one party or the other. The following quotation from *Chimpanzee Politics* illustrates how Luit, within weeks of attaining alpha rank, adopted the control role.

> On one occasion, a quarrel between Mama and Spin got out of hand and ended in biting and fighting. Numerous apes rushed up to the two warring females and joined in the fray. A huge knot of fighting, screaming apes rolled around in the sand, until Luit lept in and literally beat them apart. He did not choose sides in the conflict, like the others; instead anyone who continued to fight received a blow from him. I had never seen him act so impressively before.[35]

What is special about the control role is, first of all, its impartiality. Because my students and I recorded thousands of interventions in Arnhem, we were able to compare them statistically with partner preferences expressed in grooming and association. While most group members favored their relatives, friends, and allies when interfering in a dispute, control males were exceptional in that they placed themselves *above* the conflicting parties.[36] They seemed to interfere on the basis of how best to restore peace rather than how best to help friends. Because control interventions do not seem to be a way of pulling rank, or of helping parties who might return the favor, the chief objective seems to be to get rid of internal disturbances. The same attitude was observed by Boehm in wild chimpanzees:

After successfully damping the conflict, often an individual acting in the pacifying control role withdraws and sits down, or else returns to a prior activity such as feeding, but appears to be ready to intervene again should the conflict resume, while fights are regularly headed off, stopped, or ameliorated when these methods come into play.[37]

The second characteristic of the control role concerns the point that return favors do not seem to be the object. The quickest way to cement profitable alliances obviously would be to support the most powerful players in the group, hence the usual winners of fights. Instead, the control male often interferes in the smallest squabbles among juveniles to protect the younger of the two. In fights among adults, he supports the loser, such as a female attacked by a male—even if this male happens to be his best buddy. He does not seem to be out to please those individuals capable of doing him the greatest return favors. Indeed, a male who operated on this basis would be ineffective as arbitrator and encounter strong resistance.

We observed such resistance when Nikkie, with Yeroen's help, rose to the top in the Arnhem colony. Young and inexperienced, Nikkie favored Yeroen and certain high-ranking females whenever he interceded. Perhaps for this reason, and because of his heavy-handedness, the females did not take kindly to Nikkie's interventions and often mounted coalitions to keep him away from a scene of conflict. Yeroen's interventions, on the other hand, were impartial, controlled, and readily accepted. Within a few months, Nikkie left all control activities to the old male.

I learned from this development not only that the control role can be performed by the second-in-command, but also that the group has a say in who performs it and how. It appears to be one of those social contracts in which one party is constrained by the expectations of the others. As a result, the control role may be regarded as an umbrella shielding the weak against the strong, yet one upheld by the community as a whole. It is as if the group looks for the most effective arbitrator in its midst, then throws its weight behind this individual to give him a broad base of support for guaranteeing peace and order. In this sense, then, there is a payoff to the control role: instead of coming from powerful friends, the return favor comes from those who at first glance appear powerless. Indeed, males in this role may benefit tremendously from grass-roots support if their position comes under fire from the "angry young men" of their species.

Another advantage for the control male is that by supporting the weaker of two combatants he may be leveling the hierarchy underneath him, thus creating a bigger gap between himself and potential rivals. His interventions frustrate attempts of other high-ranking individuals to enforce their positions. Not that such a tactic can explain all control activities observed—control males often stop aggressors who cannot possibly pose a threat to their position, as when they intervene in a quarrel between juveniles—but it is another benefit to consider.

The dissociation between a control male's social preferences and the way he manages intragroup conflict is unique, showing the first signs of equity and justice. A fair leader is hard to come by, hence it is in the community's interest to keep him in power as long as possible. Tensions among mothers, for example, flare whenever a friendly rough-and-tumble among juveniles turns into hair pulling, kicking, and screaming. Because each mother is inclined to protect her own offspring, her approach to the scene is enough to raise the hairs—literally—on the other mother. A higher authority who takes care of these problems with impartiality and minimum force must be a relief for all.

Even though we cannot go so far as to call chimpanzees egalitarian, the species has certainly moved away from despotism toward a social arrangement with room for sharing, tolerance, and alliances from below. Although high-ranking individuals have disproportionate privileges and influence, dominance also depends to some degree on acceptance from below.

An example of accepted dominance is the position of the current alpha male in our chimpanzee group at the Yerkes Field Station. After removal of the previous alpha male, we introduced two adult males. The females, who had lived together for many years, ejected both of them (that is, the males had to be treated for serious injuries). Several months later, two new males were introduced. One got the same reception, but the other male, Jimoh, was permitted to stay. Within minutes of his introduction, two older females contacted and briefly groomed Jimoh, after which one of them fiercely defended him against an attack by the alpha female.

Years later, during a background check on all the chimpanzees in our study, it was discovered that Jimoh had been housed with those same two females at another institution before coming to the Yerkes center. Apparently this contact more than fourteen years earlier had made all the difference.

Although nearly thirty years old, Jimoh is an unusually small male—considerably smaller than several of the adult females he has come to dominate through sheer energy and persistence. Every day he would make charging displays, asserting himself, sometimes supported by his female friends, but increasingly on his own. One by one the females began to bow and pant-grunt to him, including eventually the alpha female. Jimoh also took up the control role: I have known few males so alert at stopping the most minor squabbles before they get out of hand.

Only on rare occasions (such as his attack on Socko, described earlier) do we see the former female alliance against him. These incidents, in which Jimoh is invariably defeated, confirm that he is dominant only by virtue of the females' acceptance, a power balance reminiscent of the first-among-equals arrangement of human egalitarian societies. However atypical this undersized male's relation with the females may be, it illustrates that dominance is not always a matter of one party's imposing conditions on others: rank relationships resemble a mutual contract. In the wild, too, the stability of a male's position may depend on acceptance of his leadership.[38]

In short, the dismantling of despotic hierarchies in the course of hominoid evolution brought an emphasis on leadership rather than dominance, and made the privileges of high status contingent upon services to the community (such as effective conflict management). Community concern comes into play here. Boehm believes that arbitration in disputes helped keep communities numerically strong by reducing divisiveness, hence allowed them to stand up more effectively against hostile neighbors.[39] This trait became so important during human evolution that it gave rise to a far more anticipatory approach to controlling conflict. Thus, in addition to breaking up ongoing disputes, or reconciling afterward—as seen in our closest relatives—people make an explicit effort to instill community values in each member, and to lay down general principles for the resolution of conflict. Probably this development required every sector of society, from the highest to the lowest, to have some say in community affairs. Rather than being some irrelevant curve in our evolutionary path, the egalitarian ethos may have been a prerequisite for the origin of morality.

QUID PRO QUO

> The fine adjustment of social relations,
> always a matter of importance among
> primates, becomes doubly important
> in a social system that involves food
> exchange. Food sharing . . . probably
> played an important part in the devel-
> opment of reciprocal social obligations
> that characterize all human societies
> known about.
>
> *Glynn Isaac*[1]

Who other than a crazed scientist would drag eight tons of dead livestock through the snow up a mountain just to see scavengers feast on the meat? Bernd Heinrich did just that in the Maine wilderness, conducting a series of experiments on the discovery of carrion by ravens and their communication about it. Ravens are about the smartest birds in the world, careful to check out any carcass before landing near it. After all, it could be a wounded or sleeping animal, or guarded by other interested parties. Thus, even the most conspicuous food bonanzas may go untouched for weeks, leaving the freezing zoologist with lots of time to fine-tune his theories.

In his enjoyable *Ravens in Winter* Heinrich notes that there are always a few ravens brave enough, despite their extreme caution, eventually to face the risks. Others will watch from the sidelines and join in only when the coast is clear. While the advantage of that attitude goes without saying, one wonders about the motivation of the birds testing the ground. Had they waited a bit longer, perhaps another raven would have done the job for them. Heinrich points out, however, that not all ravens hesitate to be the first: some seem to positively *court* danger. For example, ravens have been seen diving at a resting wolf, or sneaking up from behind to peck at its tail, only to jump away at the last moment to escape its snapping jaws.

I recognize this behavior all too well as I used to have jackdaws, small members of the same corvid family. On strolls with me through a park, a male named Johan made a habit of pestering dogs by flying just above their heads, so as to make them go after him. The poor wingless creatures would run and jump up each time Johan came close, until they were exhausted. I can only guess at the dogs' feelings, but I am familiar enough with human faces to know that the owners—especially of the bigger dogs—were not amused to see the bird return to my shoulder afterward.

Heinrich speculates that the occasional boldness of corvids serves to enhance status and impress potential mates by demonstrating that they have the courage, experience, and quickness of reaction to deal with life's dangers. Male birds may alternate between acts of bravery and courtship of female onlookers. They sometimes even treat risk-taking as a privilege rather than as anathema: tested with stuffed owls, male crows have been seen *competing* over who is going to attack the predator. In our own society, bravery may be rewarded with medals; in corvid society, according to Heinrich, it helps females tell the "men" from the "boys."

Ravens who locate a carcass, check it out, and communicate their discovery to the valley and beyond with loud yells are doing their fellow corvids a tremendous favor at a price to themselves. Without any positive impact on reproduction, bravery would long ago have been weeded out by natural selection. Wolves have been known to kill annoying ravens.

The rise of sociobiology is closely tied to successful attempts to account for the evolution of *altruistic behavior,* defined as actions that are at once costly to the performer and beneficial to the recipient. Cost-benefit analyses are the staple of evolutionary arguments. The premise is always that there must be something in it for the performer, if not immediately then at least in the long run, and if not for him then at least for his relatives. Heinrich follows this assumption when speculating that acts of courage may make male ravens attractive to the opposite sex. Showing off desirable qualities, they are rewarded by increased mating opportunities.

Because of the constantly changing composition of the feeding crowds, and of the raven population in general, it is almost impossible for these birds to achieve much more than such instant payoffs. Their encounters follow the laws of the singles-bar, not those of the marketplace. Other species are more amenable to a system in which favors are traded back and forth. Primate groups have a limited membership and tend to hold together over long periods of time. Durable relationships allow close monitoring of the exchange of favors. These animals do not need to operate on the basis of quick impressions of one another; they review entire histories of interaction.

The Less-than-Golden Rule

The idea of exchange is central to the theory of *reciprocal altruism,* developed by Robert Trivers in the early 1970s. It is quite different from the theory of *kin selection,* proposed a few years earlier by William Hamilton. The two theories are by no means in conflict, however; they complement each another nicely. Hamilton's theory explains why animals often assist close relatives. Even if such altruism harms the actor's chances for reproduction, it does not necessarily prevent the genes involved from reaching the next generation. Kindred are by definition genetically similar, sometimes even identical, to the actor and will thus put the help received to good use by spreading the same genes. From a genetic perspective, helping kin is helping oneself.

Initially, kin selection completely overshadowed reciprocal altruism in discussions about the origin of altruism and mutual aid. The focus at the time was on social insects (which often live in colonies of close relatives), and there was a dearth of examples to support Trivers' theory. In recent years reciprocal altruism has begun to receive the attention it deserves, particularly in research on primates and other mammals. This is partly owing to the current fascination with cognitive abilities; reciprocal altruism is a complex mechanism based on the remembrance of favors given and received. Even more important from a moral perspective is that reciprocal altruism allows cooperative networks to expand beyond kinship ties.

Not that support of kin is morally irrelevant—family obligations weigh heavily in any moral scheme—yet the tendency is quite robust in and of itself. Sometimes it even leads to distinctly unfair situations, as when a power figure hands out government jobs to his relatives. In human society, a bias for kin is both assumed and a matter of concern.

Our strongest moral approval is reserved for the more fragile tendencies that underlie collective welfare, such as sharing and cooperation outside the family or clan. The most effective way to instill these tendencies is through some sort of linkage between taking and giving. The usual argument is that we are better off helping each other, that the benefits of such a system far outweigh the costs of contributing to it. This anticipation of gain is central to the human moral contract—not in each and every exchange, of course, but overall. This is why no one can withdraw from the contract without dire consequences, such as ostracism, imprisonment, or execution. All able-bodied men and women are in it together. What would be the point of contributing to a community in which others idly stand by?

Reciprocity can exist without morality; there can be no morality without reciprocity. If we accept this thesis, it is clear why the very first step in the direction of the Golden Rule was made by creatures who began following the reciprocity rule "Do as the other did, and expect the other to do as you did."[2]

Even though phrased here as straightforward tit-for-tat—whereas reciprocal altruism is quite a bit more flexible and variable—the first hints of moral obligation and indebtedness are already recognizable.

Mobile Meals

A link between morality and reciprocity is nowhere as evident as in the distribution of resources, such as the sharing of food. To invite others for dinner—whether around a table or at a campsite fire—and to have the invitation returned on a later date is a universally understood human ritual of hospitality and friendship. It is difficult even to imagine a culture in which people believe that joint meals are to be avoided and that feasts are a bizarre habit. Inasmuch as such a concept would fly in the face of the most elementary solidarity, a culture embracing an eat-alone principle would strike us as a collection of individuals unworthy of the label "community," let alone "moral community."[3]

Rather more typical of our species is Japanese drinking etiquette, according to which one insults one's company by pouring oneself a drink. Friends around a table are supposed to be each other's hosts and guests at the same time: they constantly keep an eye on the glasses around them to fill them with beer or sake whenever necessary. No one goes thirsty. This custom epitomizes what human community is all about: reciprocity and sharing (see page 137).

HELP FROM A FRIEND

For food-sharing primates, the arrival of food is a joyous event that leads to begging and a relatively equal distribution. Chimpanzees and capuchins are exceptions in this respect: very few other primates show high tolerance for sharing outside the mother-offspring unit. A more widespread form of mutual aid is coalition formation: most primates will support others in fights.

Chimpanzee politics are based on coalition formation. Here two adult males are united during a confrontation. Nikkie (behind) became alpha male with the assistance of Yeroen, an older male. Nikkie mounts Yeroen, while both scream at their common rival. (*Arnhem Zoo*)

When food arrives, chimpanzees engage in a celebration. They embrace each other, hooting loudly—here an adult male—drawing the entire colony into the festive atmosphere. Blackberry shoots, a favorite foliage, are eaten with withdrawn lips because of the spines (facing page). *(Yerkes Field Station)*

Gwinnie (left) extending an open hand to Mai, who is chewing a tasty morsel. Since Gwinnie does not share much herself, others have little to gain from sharing with her. Gwinnie's begging often meets with resistance. (*Yerkes Field Station*)

A cluster of four sharing chimpanzees and an infant (partially hidden beneath their food). The female in the top right corner is the possessor (as well as the infant's mother). The female in the lower left corner is tentatively reaching out for the first time; whether or not she can feed will depend on the possessor's reaction. *(Yerkes Field Station)*

Not all food is distributed by means of passive tolerance. Here an adult female feeds an unrelated juvenile a piece of sugarcane. *(Yerkes Field Station)*

Food sharing is a highly selective process: only half of the interactions result in an actual transfer. *(Yerkes Field Station)*

Chimpanzees look through a window to see what their lucky group mates may have discovered indoors during one of our food tests. During these tests, possessors (inside) sometimes hand food through the window to the members of the colony waiting outside. *(Yerkes Field Station)*

An adult male chimpanzee in the African rain forest eating a red colobus monkey. Capture of such agile prey requires close cooperation and task division among chimpanzee hunters. *(Courtesy of Christophe Boesch; Taï National Park, Ivory Coast)*

Hunting-related communication and food sharing are less developed in baboons than in chimpanzees. After a female savanna baboon encountered a gazelle fawn, she sat down next to it, looking repeatedly at her troop in the distance. The troop included adult males who no doubt would have killed the fawn had they known about it. The female left without doing it any harm. (*Gilgil, Kenya*)

One of the many traits capuchins have in common with chimpanzees is that they share food. A juvenile displays the typical hand-cupping of a begging capuchin. He presses his face close to the food of an adult male, proffering an open hand. *(Yerkes Primate Center)*

Despite their tolerant temperament, stump-tails do not allow subordinates to take food directly from their hands or mouth. While the alpha male feeds, everyone has come to watch, but no one is getting anything. The species also lacks begging gestures. *(Wisconsin Primate Center)*

Rhesus monkeys have a talent for aggressive cooperation. They form alliances mostly to advance their position within the group, but may also do so against other species—here a dog. First an adult male threatens the dog on his own. When there is no apparent effect, he recruits a friend and together they rout the dog. *(Wisconsin farm group)*

Three rhesus females of the third-ranking matriline oppose a single member of the number two matriline (viewed from back). Although barely an adult, this individual confidently holds her ground, returning the threats. Were she to get into trouble, her entire matriline would quickly descend on her opponents. *(Wisconsin Primate Center)*

Baboon friendships often translate into support between a male and a female. A resident male baboon (middle) defends a female against a stranger (left). The stranger, who recently immigrated into the troop, first chased the female, who ran straight to her protector, hiding behind his back while screaming at her attacker. *(Gilgil, Kenya)*

Three adult male yellow baboons engage in a tense confrontation. The male on the left screams in response to the threatening stares, with raised eyebrows, of the other two. The male in the foreground used to be the ally of the threatened male but has recently begun to shift his allegiance to the male in the background. The screaming male is thus faced with a rather unpleasant change in political reality. *(Courtesy of Ronald Noë; Amboseli National Park, Kenya)*

The etiquette of sharing is universal. Here its logic is discussed in an American television classic, *The Honeymooners*. Ralph Kramden, Ed Norton, and their wives have decided to share an apartment. The men have problems with food distribution:

Ralph: When she put two potatoes on the table, one big one and one small one, you immediately took the big one without asking what I wanted.

Norton: What would you have done?

Ralph: I would have taken the small one, of course.

Norton: You would? (in disbelief)

Ralph: Yes, I would!

Norton: So, what are you complaining about? You GOT the little one!

Sharing can be so prevalent, especially in the egalitarian cultures we have considered, that the thought of keeping anything for oneself does not even arise. Katharine Milton, an American anthropologist and primatologist, tells us about this trait in Indians of the Brazilian forest:

> Unlike our economic system, in which each person typically tries to secure and control as large a share of the available resources as possible, the hunter-gatherer economic system rests on a set of highly formalized expectations regarding cooperation and sharing. This does not mean hunter-gatherers do not compete with one another for prestige, sexual partners, and the like. But individuals do not amass surplus. For instance, no hunter fortunate enough to kill a large game animal assumes that all this food is his or belongs only to his immediate family.[4]

Because hunting success is subject to chance, the man who brings home the bacon one day may be glad to get a piece of someone else's a couple of days later. Sharing makes for a more reliable food supply. Moreover, like the ravens who impress each other with their bravery, the hunter who consistently contributes more meat than his fellows may gain prestige and sexual privileges. In a Paraguayan hunter-gatherer culture studied by Hillard Kaplan and Kim Hill, for instance, successful hunters were reported to have more than their share of extramarital affairs. The anthropologists speculate that women may have sex with these men to encourage them to stay in the band.[5]

Egalitarian people being who they are, one also expects counter-measures to prevent top hunters from rising above the rest. One way is to minimize the value of whatever they have killed, even the fattest buffalo. Knowing this penchant of their comrades, the wise hunter does not boast. According to Richard Lee, the !Kung San hunter comes home without a word, sits down at the fire, and waits for someone to come along and ask what he has seen that day. He casually replies something like: "Ah, I'm no good for hunting. I saw nothing at all (pause) . . . just a little tiny one."

The anthropologist's informant adds that such words make him smile to himself as they mean that the speaker has killed something big. Yet even those who then go out with the hunter to carve up the meat will restrain their joy. Upon arrival they will cry out, "You mean to say that you have dragged us all the way out here in order to make us cart home your pile of bones!" or "To think that I gave up a nice day in the shade for this!" Then, with much hilarity, they butcher the animal, carry the chunks of meat back to camp, and the entire band eats to its heart's content.

Another informant explains why it is necessary to ridicule the hunter:

> When a young man kills much meat he comes to think of himself as chief or big man, and he thinks of the rest of us as his servants or inferiors. We can't accept this. We refuse one who boasts, for someday his pride will make him kill somebody. So we always speak of his meat as worthless. This way we cool his heart and make him gentle.[6]

In this way hunter-gatherers maintain a delicate balance between honoring contributions to the food supply—if not with their mouths, then at least with their stomachs—and subverting personal pretensions. Their entire system of reciprocal exchange would collapse if major inequalities in status were allowed to emerge.

It is evident that meat has a unique significance. Even if we no longer assume that the subsistence of hunter-gatherers owes more to hunting than to gathering, it cannot be denied that hunting products do arouse great excitement. In proportion to its nutritional value, the importance attached to meat is actually quite excessive—and universally so. Carving the Thanksgiving turkey in front of the family remains part of a North American culture that otherwise seems to be moving away from collective meals. In most parts of the world, meat (or fish) is the traditional centerpiece at the dinner table, and it is the

Animal protein takes center stage at collective meals all over the world. The division of meat is so much part of the festivity that people insist on carving it at the dinner table, as shown here for the traditional Thanksgiving turkey of North American culture. (Photo courtesy of the Norman Rockwell Museum at Stockbridge; printed by permission of the Norman Rockwell Family Trust, copyright © 1943 the Norman Rockwell Family Trust)

kind of food most often subject to special privileges and religious taboos.

Apart from cultural explanations, which stress our kinship to the animals we consume,[7] it is not hard to find a very mundane reason for the status of meat. Meat differs from vegetables and fruits in that it *moved* before it was put into the cooking pot. Meat runs, swims, or flies away if approached with unfriendly intentions. And meat that comes in large quantities tends to defend itself with teeth, tusks, hooves, and horns; it is often too much for a single hunter to handle.

Could the dangerous, unpredictable, and often cooperative nature of hunting, as opposed to gathering, explain the significance attached to meat? If so, the issue would turn from nutritional to social.

Consider an enigma that has intrigued me for a very long time: chimpanzees are quite good about sharing prey, but hopeless when it comes to favored plant foods. In the wild, the capture of a monkey is a celebrated event, signaled with a characteristic call that attracts other chimpanzees. With a deafening outburst of screams, clusters of begging and sharing individuals gather and hand meat from one to another. Why do these apes not behave the same way when they find a pile of bananas? Excitement about this food leads to such violence that provisioning at Gombe's banana camp needed to be cut back and regulated so as to reduce competition. Documenting events at the camp, William McGrew did not notice much tolerance; sharing was mostly restricted to mothers and their youngest offspring.

Here we have two favorite foods: one consists of a number of bite-size snacks, the other comes as a single large piece. The first is fought over, the second shared. One possibility is that possessors of bananas are more often attacked because a quantity of fruits is easily divided (the owner may drop half and run off with the rest), whereas an animal carcass does not fall apart when one is in a hurry to escape. Could it be, then, that aggression toward meat possessors is rare because it is not worthwhile? This explanation only takes care of the violence, however: it fails to tell us why a highly prized food that is hard to divide is shared *at all*.

It may be more fruitful to take the perspective of the hunter. When chasing arboreal monkeys, chimpanzees often work in pairs, trios, or larger teams. It is obviously no simple task to capture agile targets in three-dimensional space. Since certain species, such as colobus monkeys, defend themselves fiercely to the point of injuring the hunters, there are additional reasons for chimpanzees not to try hunting them on their own. At Gombe lone males have been seen to remain quietly near a colobus troop, only to hunt immediately when joined by another male. In Taï National Park, Ivory Coast, the hunting technique of chimpanzees has been documented by a Swiss couple, Christophe and Hedwige Boesch. Their data indicate that capture success improves with the number of participating hunters, and that the speed of movement increases sharply when the magic number of three hunters is reached.

Now, why should chimpanzee B join chimpanzee A on a strenuous hunt if A's success would not translate into any reward for B? Under such conditions B might still join A, but not in order to do anything that gives A an undue advantage; B's primary goal would be to exploit A's effect on the prey and to catch something before A does. At best, the result would be synchronous action.

According to the Boesches, chimpanzees are much better coordinated than this: they arrive at a task division in which individual hunters perform different but complementary actions. Some of them drive the prey, others encircle them or block their escape to a distant tree. If meat is the incentive for working together, Taï chimpanzees should also share readily. Such indeed proved to be the case, particularly with regard to adult males. Rather than taking meat into the trees, where beggars can be avoided, Taï chimpanzees typically form feeding clusters on the ground, where there is room for everyone.

Meat sharing is not an entirely peaceful process. The captor usually manages to retain a significant portion, yet prey does tend to change hands immediately upon the hunt's completion. There is great commotion when a monkey has been caught; others may steal the prize during the flurry of excitement that follows. All the same, participation in a successful hunt usually leads to a payoff, whether directly from the captor's hands or from someone else's. When the entire group has settled down, the begging, whimpering, and waiting for scraps begins. Soon everyone is chewing meat and crunching bones.

Chimpanzees seem to recognize a hunter's contribution to this process. The Boesches found that participation in a hunt affects how much meat a male can expect at the end. Males attracted to the scene *after* the prey is captured, whatever their rank or age, tend to receive little or nothing. Females are not held to the same standard: they receive meat regardless of their role in the hunt. According to the primatologists, the linkage for male hunters between participation and payoff guarantees a high degree of collaboration in the Taï community.

At an international conference on chimpanzee behavior organized by the Chicago Academy of Sciences in 1991, two Japanese primatologists, Hiroyuki Takasaki and Toshisada Nishida, showed a videotape in which a male chimpanzee seemed to reward a female directly for assisting him in a hunt. Here is my summary of the sequence.

After an adult male had captured one monkey, a female discovered a second, which she cornered underneath some rocks. Instead of trying to catch it herself, she called the male while staring at the monkey. The male understood her signals, came over, and grabbed the second monkey. He killed it right away by flinging it forcefully against the ground. Despite the presence of other beggars, the male let this particular female have one of his two monkey carcasses.

Contrast this with an encounter I once witnessed between a female baboon and a Thompson gazelle fawn hidden in the high grass of a Kenyan plain. Even though baboons are known to eat fawns, the female merely stared at it. My guide, Ronald Noë, who has studied these baboons, suggested that the fawn was too large for her to kill; an adult male would not have hesitated. The female was more than 30 meters away from her troop. She looked around at the others several times, but did not vocalize or in any other way draw attention to the situation.

The fawn was lucky. Had food-related tolerance been as well developed in baboons as in chimpanzees, the female could have called in a male in the same way her chimpanzee counterpart did in the videotaped incident. Baboons do not share nearly as well as chimpanzees, however; the female would not have gotten any.

The orchestrated hunt of chimpanzees is unique among nonhuman primates. There are indications that their close collaboration, and the sharing that follows, also has profound implications for other aspects of social life. The Boesches note that generosity marks males of high status, and Nishida and his coworkers report on the specific case of an alpha male named Ntologi. This male appeared to use meat distribution to strengthen his power base in the Mahale Mountains community. Apart from being an effective hunter himself, Ntologi often claimed carcasses from other males: in about one-third of all meat-sharing sessions, he controlled the largest portion. To distribute food appeared to be his main intention; he occasionally secured carcasses from which he scarcely ate himself. He would merely hold the spoils and let others pull off meat until everything was gone.

There was a peculiar system to Ntologi's food distribution. When plundering a carcass from prime adult males, he always permitted these males to keep a substantial cut for themselves. Young adult males did not receive this courtesy, nor were they allowed to take scraps afterward. With Ntologi in control, males less than twenty years of age could do little other than watch from a distance while the rest feasted. Around the age of twenty, males are climbing the social ladder and begin to pose a serious challenge to high-ranking males. Similarly, Ntologi never permitted beta males, the most likely usurpers of the alpha rank, to join the meat-sharing cluster.

Kalunde became the beta male in 1986 and since then has become Ntologi's major rival. Although he is usually observed not far from Ntologi, Kalunde typically disappears from the scene during meat-eating episodes when Ntologi controls access to the carcass. Kalunde has never been seen to beg for meat from Ntologi. The former alpha male did not share meat with Ntologi when he was the beta male.[8]

Ntologi chiefly shared with females, as well as with those males who were unlikely to threaten his position; that is, influential aging males and prime adult males in stable midranking positions. While these males lacked the social status and/or physical capacity to rise to the top, they could be effective allies. Some were seen to join Ntologi's intimidation displays against rivals. Possibly the alpha male was "bribing" certain parties to support his position, or at least not turn against him. Ntologi's tenure at the top of the Mahale community lasted an exceptionally long time—more than a decade. Perhaps the way he distributed meat was part of the secret.

During one of the power struggles at the Arnhem Zoo, the challenging male, Luit, developed a sudden interest in live oak and beech trees on the island. Their foliage, much liked by the chimpanzees, was almost impossible to reach as every tree trunk was surrounded by electric wire. Luit would break a long branch from a dead tree, then drag it to a live tree to use it as a ladder to avoid being shocked—an act of great skill and bravery. Once high in the live tree, he would break off many more branches than needed for himself, only to drop them to the colony waiting below. Soon everyone, including rival males, would be munching on fresh leaves from the sky.

Because Luit accomplished this feat many times during his successful bid for alpha status, I speculated in *Chimpanzee Politics* that "it may have been accidental that Luit played Santa Claus on that first and later occasions, strewing food around for the group, but to me it immediately looked as if he had hit on a very clever way of drawing attention to himself."[9]

Possibly, then, generosity serves political ends: food distribution may enhance an individual's popularity and status. If so, the ancient top-to-bottom arrangement remains visible in the more recently acquired economy of exchange. Yet instead of dominants standing out because of what they take, they now affirm their position by what they give. Note that this is precisely the sort of status seeking that human egalitarian societies try to prevent.[10]

Early human evolution, before the advent of agriculture, must have been marked by a gradual loosening of the hierarchy. Food sharing was a milestone in this development: it both marked the reduced significance of social dominance and provided a launching pad for further leveling. It is therefore relevant for us to understand its origin. In a straightforward rank order, in which dominants take food from subordinates, the food flow is unidirectional. In a sharing system, food flows in all directions, including downward. The result is the relatively equitable distribution of resources that our sense of justice and fairness requires.

Reality may not always match the ideal, but the ideal would never even have arisen without this heritage of give-and-take.

At the Circle's Center

In the animal kingdom as a whole, sharing is most prevalent in species that feed on high-energy foods of which the collection, processing, or capture depends on special skills or rare opportunities. Among non-primates, sharing is most notable in social carnivores, such as wolves, brown hyenas, mongooses, and vampire bats. In the primate order, we find it in two forms. In small, stable family units, such as those of gibbons and marmosets, food is freely shared with offspring and mates. Because the tendency benefits kin and reproductive partners, it is easily explained as a product of kin selection. Such intrafamilial tolerance may have laid the groundwork for the second kind, which could not have evolved without some degree of reciprocity: the group-wide sharing of chimpanzees, humans, and capuchin monkeys.

Certain food characteristics are conducive to this transition from kin-based to extrafamilial sharing. Whereas a few plant foods, such as giant fruits that are hard to open, show some of these charac-teristics, only large prey shows them all. Shared food typically has the following traits:

1. Highly valued, concentrated, but prone to decay.
2. Too much for a single individual to consume.

3. Unpredictably available.
4. Procured through skills and strengths that make certain classes of individuals dependent on others for access.
5. Most effectively procured through collaboration.

A hunting connection for the evolution of sharing seems obvious enough in chimpanzees and humans.[11] What about capuchins? Until recently, the sharing habits of this Neotropical monkey were relatively unknown to science. Remarkable tolerance had been reported by field workers such as Charles Janson, but mainly between mother and offspring or adult males and juveniles. My own observations of captive capuchins, however, showed tolerance to be far more pervasive. The capuchin is one of very few primates in which unrelated adults, including those of the same sex, peacefully remove food from each other's hands.

Capuchin monkeys exploit a great variety of plant and animal foods in their natural habitat. They do so through force, destructive manipulation, and perhaps tool use. I once followed a group by ear through dense forest: they were noisily breaking open palm fronds. Because of the strength involved, this foraging technique may itself provide a basis for sharing. Adult males (who are very muscular compared to females) may be able to reach foods that females and youngsters cannot obtain on their own.

Most animals eaten by these omnivores are insects, spiders, reptiles, tree frogs, and so on. Not all prey are too tiny to share, however. Susan Perry and Lisa Rose, anthropology graduate students, recently discovered that wild capuchins in Costa Rica raid coati nests and distribute the meat of the pups. Apart from raids on unguarded nests, the monkeys also managed to steal pups defended by adult coatis. Even for agile monkeys this is a risky business, as coatis have formidable jaws. Coatis are relatives of raccoons, roughly twice the size of capuchins.

While the degree of cooperation—if any—during these raids remains to be ascertained, subsequent sharing has been carefully documented. Because the capuchins ate the pups alive, without killing them first, beggars were attracted by the screams of the prey. Each carcass had between one and seven owners before it was consumed. Most of the sharing took the form of owners allowing others to collect scraps or tear off pieces. On occasion, however, they would set meat down in front of a beggar. The capuchin monkey seems no exception, therefore, to a connection between predation and food division.

If carnivory was indeed the catalyst for the evolution of sharing, it is hard to escape the conclusion that human morality is steeped in animal blood. When we give money to begging strangers, ship food to starving people, or vote for measures that benefit the poor, we follow impulses shaped since the time our ancestors began to cluster around meat possessors. At the center of the original circle we find a prize hard to get but desired by many, a situation that remains essentially unchanged. In the course of human evolution and history, this small, sympathetic circle grew steadily to encompass all of humanity—if not in practice then at least in principle. Some philosophers, such as Peter Singer in *The Expanding Circle,* even feel that all creatures under the sun deserve to be included, and that animals should therefore never be used for research, entertainment, or human consumption. Given the circle's proposed origin, it is profoundly ironic that its expansion should culminate in a plea for vegetarianism.

I am not saying that such arguments cannot or should not be made. Moral reasoning follows a logic of its own, independent of how the human inclinations that it works with (or against) came into existence. The moral philosopher simply takes for granted what the biologist seeks to explain.

A Concept of Giving

Ever since the meat sharing of chimpanzees was dismissed as "tolerated scrounging" by Glynn Isaac, a paleontologist interested in the role of hunting and sharing in human evolution, anthropologists have felt absolved from the need to take its implications seriously. According to Isaac, only active food distribution deserves the label "sharing."

It is indeed highly unusual to see one chimpanzee spontaneously offer food to another. Yet, instead of stressing what these apes do not do, we may find it more profitable to focus attention on what they do do in comparison with most other species. Many primates do not seem to have even a *concept* of giving. When I for the first time hold out an apple to a rhesus monkey, he will most likely snatch it out of my hand with menacing stares and grunts, the only way he knows to obtain food from someone else. Because relaxed give-and-take is not part of his natural repertoire, it takes time for a rhesus monkey to learn not to bite the hand that feeds him.[12]

Not only do chimpanzees (and other apes) react more amicably to food offerings, they understand exchange. For example, if something that we wish to retrieve (such as a screwdriver) has been left behind in the ape enclosure, one of its inhabitants will quickly grasp what we mean when we hold up a tidbit while pointing or nodding at the item. She will fetch the tool and trade it for the food. I know of only one monkey species with which such deals can be struck without prior training, and it can be no accident that it is yet another food sharer: the capuchin.

Even though food donation is normally rare, situations can be created in which it becomes common. As long as sixty years ago, Henry Nissen and Meredith Crawford successfully did so. They conducted experiments with pairs of chimpanzees in adjacent cages separated by bars; only one of the two was given food. After the other begged with hand held out, some possessors shared with them, which obviously required active participation. Similarly, we recently conducted tests at the Yerkes Field Station in which chimpanzees were trained to enter a building individually and open a box. On some occasions the box was stuffed with fruits. The rest of the colony could watch the discovery through an open window too small to climb through. Soon it became known among us as the "take-out window" because of the occasional handouts to outsiders by the fortunate ape inside.

Although active sharing constitutes only a fraction of all food transfers in the wild, the phenomenon has been reported by many students of chimpanzee behavior. Most accounts concern meat sharing, but Jane Goodall once saw an adult female bring fruit to her aging mother.

Madame Bee looked old and sick. It was very hot that summer, and food was relatively scarce. . . . When soft calls indicated that the two young females [her daughters] had arrived at the food site, Madame Bee moved a little faster; but when she got there, it seemed that she was too tired or weak to climb. She looked up at her daughters, then lay on the ground and watched as they moved about, searching for ripe fruits. After about ten minutes Little Bee climbed down. She carried one of the fruits by its stem in her mouth and had a second in one hand. As she reached the ground, Madame Bee gave a few soft grunts. Little Bee approached, also grunting, and placed the fruit from her hand on the ground beside her mother. She then sat nearby as the two females ate together.[13]

As for capuchins, the first evidence for active sharing is buried in a technical report by M. R. D'Amato and Norman Eisenstein. During experiments on food deprivation one subject, Lucy, failed to lose weight. The authors blamed this surprising result on Lucy's friends. "The most likely interpretation . . . is that Lucy received additional food supplies from unauthorized sources. We have observed more than once the passing of food from an animal on full rations to one in an adjacent cage on deprivation."[14]

Our ongoing research confirms this observation. Inspired by the Nissen and Crawford study, we place capuchin monkeys in adjacent compartments of a test chamber with a mesh partition between them. Rather than deprive them, we give them treats of attractive food.[15] One of the two receives a bowl of apple pieces for twenty minutes, after which the other receives cucumber pieces for the same amount of time. Many a colleague has been amazed by our video footage of capuchins handing, pushing, or throwing food through the mesh to their neighbor. This kind of exchange is simply unthinkable in most other primates.

Nevertheless, in these same tests the vast majority of transfers remain passive. The possessor generally transports handfuls of food

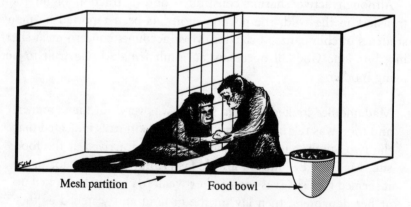

Mesh partition ⟶   Food bowl ⟶

A male capuchin (right) hands a piece of apple to a female. The mesh between the two monkeys prevents the female from getting any food without the male's cooperation. In the same test, the female will receive a bowl with cucumber pieces to see if her willingness to share is affected by what she received. This drawing (by the author) is based on an actual video still.

Science's first videotaped instance of what we interpret as spontaneous deception occurred unexpectedly during a food test on chimpanzees. It contains all the crucial elements desired in a case of deception: a clear motive, a benefit, and the absence of trial-and-error learning. The individual committing the apparent act of miscommunication had never before found herself in this exact situation.

The following description is based on an analysis of two videotapes—one recorded indoors, the other outdoors—as well as direct observation by four experienced chimp watchers. Time is in seconds from the onset of the test.

Date: May 21, 1993

Location: Yerkes Primate Center Field Station

Time   Simultaneous Indoor and Outdoor Observations

000   Inside a building, Marilyn and her juvenile daughter discover a box filled with two dozen apples. Outside, the rest of the group is aware of the food because of an open window.

040   While Marilyn and her daughter gorge themselves, Georgia, an adult female, begins to collect handfuls of mud (it has rained) in the outdoor enclosure. Sticking her arm through the window, Georgia hurls the stuff with excellent aim at the two inside. Marilyn (although quite dominant over Georgia) goes to the window with complaining whimpers.

460   The indoor area is becoming dirty, slippery, and wet. Dirt throwing is not unusual in this colony for individuals who want attention, but what Georgia is doing amounts to serious pestering. Because she does not let up, Marilyn is forced to leave the box. She climbs to the ceiling, out of Georgia's range. Marilyn's daughter does not seem to mind and continues to feed, dodging the incoming mud balls.

620   During a brief spell of quiet, Marilyn's daughter hands a piece of apple through the window to Georgia's daughter outside.

635   Marilyn comes down from the ceiling and moves to the window while uttering soft calls. These particular calls normally signify mild alarm, as for a distant snake. Marilyn does not look at anything in particular, though, and the experimenter inside sees absolutely nothing disturbing.

645   Marilyn now utters louder alarm calls through the window. Georgia immediately responds by rushing to the fence of the enclosure to stare into the forest behind the building. She then rushes away from the fence; her startled response makes other apes do the same. The apes outside gather on a climbing frame (a normal response to danger).

Even though we inspect through binoculars the spot on which Georgia focused her attention, we detect nothing unusual. Furthermore, if anything is there, Marilyn can't have seen it as it is outside her visual range (blocked by the building's side wall).

710 Still staring in the same direction, the chimpanzees outside now utter a chorus of loud "woaow" barks. Marilyn joins them by calling from inside, looking out through the window.

725 Marilyn's daughter seems disturbed by the alarm—she does not approach the box any more and stays alert—as do the others outside. Marilyn is *the only one* who pays no further heed. Immediately after her last bark, she turns away from the window and moves to a far corner of the indoor area. There she picks up an apple and starts eating.

Georgia has lost interest in annoying Marilyn, who can finally enjoy her food bonanza.

---

to the partition, spreads it out on the floor, and starts eating. The partner needs only to reach through the mesh to collect pieces, or take them directly from the possessor's hands or mouth. Typically, capuchins cup an open hand under the possessor's chin in order to catch any dropping bits. Note that there is absolutely no need for possessors to bring food to this side of their compartment; they could just as easily keep everything for themselves in a far corner. Indeed, this is what occurs in intolerant pairs. Sharing is a *selective* process; we are able to predict on the basis of social relationships in the group which combinations of individuals will share, and which will not.

If active donation is rare in species other than our own, it is not because they are incapable of it. Rather, there seems little need for it. Whether transmissions of food are brought about by possessors or recipients is of secondary importance; the main point is that transmission takes place. Letting others gently remove food that is clearly in your possession is a tremendous change compared to animals without such tolerance. Chimpanzees and capuchins may not share exactly like people, but the outcome is identical: one individual obtains food from another on an apparently voluntary basis.[16]

Testing for Reciprocity

The psychology of sharing is little understood. For example, our capuchins may carry food to the mesh partition simply because they prefer to eat next to their companion. That is, the sharing could be a combination of tolerance and affection. Alternatively, their behavior could reflect a concept of exchange and the expectation of return favors. Which of these two possibilities is the reality makes quite a difference. It is this sort of question that our experiment seeks to answer, issues difficult or impossible to resolve in the field. Now that we know *how* and *when* primates share food under natural conditions, it is time to manipulate this tendency in a controlled environment to establish *why* they do so.

The need for such research is particularly felt in relation to chimpanzees. Despite all the observations of meat sharing in the wild, virtually nothing is known about the degree of reciprocity, partly because participation in predatory episodes changes from one occasion to the next, as does the quantity of meat. Possibly the apes themselves keep track of who did what for whom, but the constantly changing composition of feeding clusters makes the establishment of reciprocity a statistical nightmare for the scientist. How much more convenient to conduct repeated standardized trials with the same individuals. This is precisely what I set out to do at the Yerkes Field Station.

What kind of food to select? I knew that a single pile of favored fruits, if provided on a regular basis, would be the surest way of destroying the colony through violence. We learned this lesson from the bananas at Gombe as well as from early attempts at establishing captive chimpanzee colonies. To allow the chimpanzees to catch and kill an animal might provide a natural context for a study of sharing, but was ruled out for ethical reasons. Instead, I gave the colony large bundles of branches tied together with honeysuckle vine.

Wild chimpanzees do not need to share the foliage that is all around them. In captivity, however, branches with fresh leaves are ideal to investigate sharing; they arouse quite a bit of excitement yet no excessive competition. When the chimpanzees see a caretaker arrive in the distance with two enormous bundles of blackberry, sweetgum, beech, and tulip tree branches, they burst out hooting. General pandemonium ensues, including a flurry of embracing and kissing. Friendly body contact increases one-hundred-fold, and status signals seventy-five-fold. Subordinates approach dominants, particularly the alpha male, to greet them with bows and pant-grunts. Paradoxically, the apes are confirming the hierarchy just before canceling it, to all intents and purposes.

I call this response a *celebration*. It marks the transition to a mode of interaction dominated by tolerance and reciprocity. Celebration serves to eliminate social tensions and thus pave the way for a relaxed feeding session.[17] Nothing even remotely similar occurs in species that do not share. If macaques notice the arrival of attractive food, they immediately move into a competitive mode: high-ranking monkeys come forward, supplanting those of low rank. Chimpanzees do the exact opposite, throwing themselves into each other's arms with obvious delight. Within minutes each and every member of the colony has obtained some food. They do show competition, occasionally even fight, but it is their peacefulness and civility that is most striking: only 3 percent of interactions between adults involve any sign of aggression.

Field-workers noted long ago that social dominance loses some of its luster during sharing sessions. It does not become totally insignificant—as illustrated by Ntologi—but it is not unusual to see the most feared and respected individual stretch out a hand to one of his underlings to beg for a few scraps. Why does he not claim the food by force? In our tests, subordinates avoid dominants if both approach unclaimed food, but once the food is firmly in a subordinate's hands, his or her ownership is generally respected. This pattern holds among adults only; juveniles are treated (and treat each other) with less inhibition.[18]

The generosity of high-ranking individuals is so striking that subordinates often flock around them rather than solicit food from low-status possessors, who tend to hold onto it more tightly. This observation contradicts a popular hypothesis, according to which possessors (human or nonhuman) part with their load out of *fear*. Nicholas Blurton Jones, a British anthropologist, proposed that sharing is a way of keeping the peace with potentially hostile beggars. If true, would we not expect low-status possessors to share the most, as they are the most easily intimidated?

Walnut, the previous alpha male of the Yerkes colony, was the most generous sharer in my studies. Huge and muscular, without rivals of his own sex, Walnut totally dominated all females and juveniles. To me it is unimaginable that there was one drop of fear in this male when he let females pull the juiciest branches out of his bundle. The sharing-under-pressure hypothesis also cannot explain why chimpanzees and capuchins still bring food to others when the risk of attack has been eliminated, as in the experiments previously described. Finally, the hypothesis fails to account for the predominant direction of aggression in my food trials with chimpanzees; possessors threatened beggars five times more often than that beggars threatened possessors.

It remains possible that sharing under pressure comes into play with highly valued foods, or with a large number of competing males. Yet the distribution of food is so well developed under more relaxed conditions that I see this as a secondary development only. In non-sharing species, a single dominant may feed undisturbed even if surrounded by a hundred hungry subordinates. That such perfect peace is not given to a chimpanzee or human owner of prized food is because of different *expectations* in the have-nots, who beg doggedly and throw tantrums if ignored. What needs to be explained is the origin of these expectations, rather than the pressuring tactics that spring from them. Only the theory of reciprocal altruism provides a satisfactory answer.

To test this theory, I recorded nearly five thousand interactions over "browse food" in the Yerkes chimpanzees. Half of these resulted in a transfer of food from one individual to another—a proportion nearly identical to that determined by Geza Teleki for meat sharing in the Gombe chimpanzees. That only one of two attempts is successful shows how discriminating the apes are. The most common ways of getting rid of beggars are to turn away when they approach, to pull branches out of their reach, to tap their hand when they reach out, or just to walk off with the food. Only rarely do possessors bark or scream at beggars, or whack them over the head with an entire food bundle to make clear how little they care for their presence. Low-ranking possessors are just as capable of rebuffing high-ranking beggars as the other way around. Obviously, sharing is no free-for-all.

Food transfers in the colony were analyzed in all possible directions among adults. As predicted by the reciprocity hypothesis, the number of transfers in each direction was related to the number in the opposite direction; that is, if A shared a lot with B, B generally shared a lot with A, and if A shared little with C, C also shared little with A.[19] The reciprocity hypothesis was further supported by the finding that grooming affected subsequent sharing: A's chances for getting food from B improved if A had groomed B earlier that day.

Even though my tests were conducted on a colony with an unusual sex ratio (several adult females and a single male), and with a food source that wild chimpanzees do not need to share, the data definitively support the *principle* of reciprocal exchange. And the principle is really all that matters. Individuals with the mental capacity to keep track of given and received favors can apply this capacity to almost any situation. Collaborative hunting and the sharing of meat, or grooming and the sharing of foliage, are only two possibilities. Female chimpanzees, for example, may follow reciprocity rules when protecting or baby-sitting each other's offspring, and sex is an obvious bargaining chip between males and females. This alternative is particularly used by bonobos: females of this species are known to receive food from males immediately following, or even in the midst of, intercourse.[20]

Once a quid pro quo mindset has taken hold, the "currency" of exchange becomes secondary. Reciprocity begins to permeate all aspects of social life.

From Revenge to Justice

Tangible exchanges, such as those involving food, are often more convincing than intangible ones. Nevertheless, it is quite possible that before food sharing appeared on the evolutionary scene the trading of *social* services was already in full swing in our primate ancestors, and reciprocity had become the new way of doing business. A prime candidate for having set this process in motion is support in fights, also known as alliance formation.

As with the sharing of food, the importance of alliances is best illustrated by contrasting species that form them with species that do not. *Felis silvestris catus,* the domestic cat, belongs to the latter category. At home we have two neutered tomcats, Diego and Freddie. Diego used to roam and defend a large territory around the house. He did so effectively and diligently until he got older; then he began to withdraw (in the typical slow-motion fashion of cats) from younger and stronger rivals. Freddie, on the other hand, habitually overestimates his fighting abilities. He regularly loses against feral tomcats and does not seem to learn anything at all from these experiences.

One day Freddie, with all hair on end, stood opposite a rival who had emerged from the woods. When Diego joined him to stare at the same intruder, this looked to us like a promising approach. But when Diego took an additional step forward, Freddie immediately turned around and hissed at him, as if he was now dealing with two rivals at once. The tension of the moment had made him hostile to a cat with whom he normally was friendly. During the confusion, the enemy took off.

Because of the millions of years of alliance formation in my evolutionary heritage, I can wrap an arm around a friend's shoulder while hurling nasty insults at someone else. My friend will have no trouble appreciating that these insults have nothing to do with him. Such distinctions come naturally. Teaming up against a rival requires moving from general motivational states to differentiated states, depending on the precise role of the other. Freddie was unable to separate friend from foe, but many primates are masters at juggling attitudes. They (and a few other animals, such as dolphins) are perfectly at ease with the role differentiation of political animals. It is not that unusual for a chimpanzee to beg for A's support (with outstretched hand), while threatening B (with screams and barks), only to court C (with erect penis and foot stamping) immediately afterward.

The reciprocity of alliances is still very much under debate. Many alliances are kin based—witness the confrontations between matrilines in macaques and baboons. The evolution of such a support system does not require reciprocity; kin selection offers an excellent explanation. For nonrelatives to support one another, however, we assume that they must gain something by doing so.

Sometimes the act of support has an immediate payoff and so may not be truly altruistic. Take the way two male savanna baboons together displace a rival from a sexually receptive female. Altruism implies absorption of a cost for the sake of someone else, yet each male in these situations seems to be out for himself. Their goal is to claim the female at the expense of all other males present, including their allies. Once the previous owner has been defeated, both allies race to the female. The first to arrive will become her next consort. Sometimes the previous owner continues to do battle with one ally while the other sneaks out to the female. Clearly, "opportunism" better characterizes such behavior than "altruism." In independent studies in Kenya, both Frederick Bercovitch and Ronald Noë concluded that willingness to assist other males is not what drives these alliances. Rather, there are immediate advantages to be gained.

Not that there are no mutual benefits. There are males who operate consistently as a team; they would obviously not be doing so if, on balance, both of them would not gain. Whereas supportive relationships among male baboons may start out on an entirely opportunistic basis, some of them may gradually develop into well-established partnerships regulated by expectations of reciprocity.

Baboon males also support females and their offspring. By developing friendships with particular members of the opposite sex, males may win acceptance in a troop (male baboons migrate, whereas females stay all their lives in the same troop) while at the same time increasing the willingness of their protegees to mate. So the male offers protection—which he is eminently capable of doing, as he is almost twice a female's size—and the female offers opportunities for sex and, ultimately, reproduction. According to Barbara Smuts, these mutually profitable relationships may last for years, and are established through rather humanlike flirtations.

> First, the male and female looked at one another but avoided being caught looking. This phase was always accompanied by feigned indifference to the other, indicated by concentrated interest in grooming one's own fur or by staring intently at some imaginary object in the distance (both ploys are commonly used by baboons in a wide variety of socially discomfiting situations). Eventually, the coy glances were replaced by a more direct approach, usually initiated by the male. Grooming then followed. If the friendship "took," the couple would then sit together during daily rest periods, and eventually they would coordinate subsistence activities as well.[21]

Mixed-currency exchanges, such as a friendship to which each sex makes its own unique contribution, do not make it any easier for the investigator to establish reciprocity. Monkey groups may be veritable marketplaces in which sex, support, grooming, food tolerance, warnings of danger, and all sorts of other services are being traded. To us scientists falls the task of figuring out the worth of each service and of following relationships over a long enough period to understand the deals that are being struck. To be protected by a male might mean a lot to a female baboon—it could spell the difference between life and death for her offspring—yet is it worth enough for her to ignore the attentions of other males? Can any amount of grooming compensate the groomee for injuries suffered in an attempt to defend the groomer? We know far too little to answer such questions.[22]

For the moment, it may be simpler to isolate a single currency and determine how it is being traded against itself. We have already done so for food sharing. Applying the same procedure to thousands of alliances recorded over the years in both chimpanzees and two kinds of macaques, we were able to demonstrate that alliances too are reciprocal. That is, the tendency of A to support B in fights varies with the tendency of B to support A. While considerably stronger in chimpanzees than macaques, this relationship characterizes all three studied species.

Such reciprocity can come about in two quite different ways. The simpler is that monkeys and apes preferentially aid kin and friends, that is, partners with whom they spend much time. Does Sonja support Mira because they are always near each other, and does this also explain Mira's support for Sonja? Since association time is symmetrical (identical for both Sonja and Mira), it may be all we need to explain the mutuality of their support. There are statistical ways to remove the effects of a variable, however; reciprocity was found to persist after symmetry was taken into account. Therefore the first explanation cannot be enough, and a second mechanism must be assumed. Some of the events considered in this analysis occurred weeks or months apart. Perhaps primates keep track of favors given and received, and make assistance to others conditional on the way they have been treated in the past. This more calculative view is generally accepted for people; it may hold for other species as well.[23]

We further found that chimpanzees achieve more complete reciprocity than macaques. Not only do chimpanzees assist one another mutually, they add a *system of revenge* to deal with those who oppose them. When an individual takes sides in an ongoing dispute, his action does more than benefit one party; he cannot avoid hurting another. Each *pro* choice entails a *contra* choice. If Yeroen joins a fight between Luit and Nikkie, he simultaneously makes a *pro* choice (for Nikkie) and a *contra* choice (against Luit). This phenomenon is all too familiar in human politics. As soon as a public figure takes up a particular cause or endorses a particular candidate, he is perceived—rightly—as having chosen against other causes or candidates. It is impossible to make everybody happy.

Chimpanzees are unique in that not only the *pro* but also the *contra* interventions are reciprocally distributed; that is, if A often helps others against B, B tends to do the same against A. This tendency may sound natural enough—because people do the same—but it does not apply to macaques. To square accounts with dominant adversaries requires opposing the hierarchy, which is difficult or impossible for macaques. With the vast majority of their alliances directed down the hierarchy, there is no room for a revenge system. Chimpanzees are much less reluctant to stand up against those who normally dominate them; retaliaton is an integral part of their system of reciprocity.[24]

Thus, it is not unusual for a dominant chimpanzee to corner a group mate sometime after a confrontation in which the latter, along with a number of others, stood up against him. With his allies out of sight, this individual is in for a hard time if the dominant decides to get even. As a result, low-ranking chimpanzees need to think twice before poking their noses into disputes among higher-ups. On the other hand, once a subordinate has developed an antagonistic relationship with a specific dominant, it is only a matter of time until this dominant will become involved in a conflict that can be exploited. The subordinate—perhaps together with others in a similar mood—will side with the dominant's rival and make clear that having too many enemies involves a price.

Such retribution may be rare in macaques; it is not wholly absent. An exceptional yet dramatic instance was observed by one of my students, Brenda Miller, in a colony of Japanese macaques on an island at the Milwaukee County Zoo. The group's lowest-ranking female, Shade, was frequently harassed by a particular male. One day she finally summoned the courage to counter his aggression, turning around in the midst of a chase to face her tormentor.

What happened next explains why such behavior is rare. Within seconds Shade was subdued by an angry mob: a whole group of monkeys drove her straight into the moat. It was March, when the Wisconsin weather is still icy cold. According to Miller, attackers normally do not keep victims in the water for more than a few minutes, but Shade was forced to stay for more than twenty minutes. When she finally crawled onto the shore, she was seized and bitten by five aggressors simultaneously. Subsequent veterinary inspection showed her injuries to be superficial, but it was decided to keep Shade indoors overnight because of serious hypothermia.

Early next morning, Miller was at the zoo for the release, worried about renewed violence against Shade. Rather the opposite happened. As soon as Shade appeared, the other macaques on the island began to run away, and a high level of tension was evident. Shade, still weak and limping, walked straight to the male who had led the assault on her. During the next seventeen minutes she attacked him thirteen times, completely ignoring attempts by others (including the alpha male) to stop her. Each time she was distracted, her two offspring would take up the chase after the culprit. Eventually aggression subsided. The male approached Shade three times to make contact, but she obviously was not in a forgiving mood; she moved away from him an equal number of times. The male hit her, made a bouncing display with his tail in the air, and left Shade alone. Only then did the group return to normal activities.

Joint action by low-ranking against high-ranking macaques is extremely uncommon. Perhaps to compensate for this deficit, macaques are specialists in *indirect* revenge. Filippo Aureli and his colleagues discovered that victims of attack often vent their feelings on a relative of the opponent. Their targets are typically younger than the initial aggressor, hence easier to intimidate, and the vindictive action may occur after considerable delay. It is as if I respond to a reprimand received from my boss by looking for his daughter to pull her hair. I do not need to go against the peck order, yet I still inflict punishment on my offender. Unfortunately for me, my boss will have no trouble handling the situation! Macaques, in contrast, cannot be fired; all that can happen is an alliance in defense of the younger target, resulting in revenge upon revenge. Matrilineal politics are cyclic and complex.

I am the first to admit that we still have a long way to go before we can consider reciprocal altruism proven and understood, yet there is every indication that it is better established in our closest relatives than previously thought. Inclusion of negative acts considerably broadens the scope of the balance sheets they seem to keep on social affairs: not only are beneficial actions rewarded, but there seems to be a tendency to teach a lesson to those who act negatively.

It is to Trivers' credit as a theoretician that—before we knew much about alliances and mutual aid—he recognized that a system of favors and return favors would not last if the tendencies that undermine it went unchecked. There is a strong temptation to take advantage of the system without making a corresponding investment. Defined as "cheating"—that is, giving less than one takes—this attitude threatens the entire system, including the interests of honest contributors. The only way they can protect themselves is by making cheating costly. They do so through punitive action, also known as *moralistic aggression*. It is an apt label, as the reaction concerns how others "ought" to behave.

A human emotion specifically mentioned in this context by Trivers is *indignation*. We are indignant, for example, when refused a favor by a colleague whom we have helped numerous times in the past. Indignation is aroused by the perception of injustice; as such, it is part of the emotional underpinning of human morality. The outraged reaction that it may trigger serves to clarify that altruism is not unlimited: it is bound by rules of mutual obligation.

A first, tentative suggestion that chimpanzees may know this special emotion came from our experiments on food sharing. We found that some individuals, when approaching a possessor, were more often threatened away than others, and that this depended on how they themselves acted as possessors. The contrast between two females, Gwinnie and Mai, illustrates this point. If Gwinnie obtained one of the large bundles of browse, she would take it to the top of a climbing frame, where it could easily be monopolized. Except for her offspring, few others managed to get anything. Mai, in contrast, shared readily and was typically surrounded by a cluster of beggars. Guess who met with more resistance if she herself was in need and tried to get food? Gwinnie, Georgia, and other stingy personalities encountered far more threats and protestations than generous sharers such as Mai and Walnut.

This relationship between an individual's generosity as possessor and the resistance encountered as beggar is to be expected if moralistic aggression regulates food exchanges. It is as if the other apes are telling Gwinnie, "You never share with us, why should we share with you!"

A taste of revenge is the other side of the coin of reciprocity. To act negatively toward stingy individuals (along with direct and indirect retaliation) suggests a sense of justice and fairness. The eye-for-an-eye mentality of primates may serve "educational" purposes by attaching a cost to undesirable behavior. Although the human juridical process abhors raw displays of emotion and prefers arguments based on the greater good of society, there is no denying the role of such urges. In *Wild Justice* Susan Jacoby points out that human justice is built on the transformation of a primitive sense of vengeance. When relatives of a murder victim or survivors of war atrocities seek justice, they are driven by a need for redress even though they may present their cause in more abstract terms. Jacoby believes that one measure of a civilization's complexity is the distance it creates between the aggrieved individual and the administration of his or her urge for vindication.

The struggle to contain revenge has been conducted at the highest level of moral and civic awareness attained at each stage in the development of civilization. The self-conscious nature of the effort is expectable in view of the persistent state of tension between uncontrolled vengeance as destroyer and controlled vengeance as an unavoidable component of justice.[25]

Thus, retributive justice ultimately rests on a desire to get even, a desire that needs to be reined in and subjected to powerful rationalization so as to keep society from being torn apart by eternal feuding. Like fire, revenge can be incredibly destructive if left untamed. According to the American anthropologist Napoleon Chagnon, it is hard even to imagine the terror that might invade our lives in the absence of laws prohibiting people from seeking lethal retribution when close kin have died at the hands of someone else. In a study of blood feuding among the Yanomamö Indians of the Amazonas, Chagnon estimated that approximately 30 percent of adult male deaths were due to violence.

A young Yanomamö man visiting the capital immediately saw the advantages of a more even-handed juridical apparatus:

> He excitedly told me that he had visited the governor and urged him to make law and police available to his people so that they would not have to engage any longer in their wars of revenge and have to live in constant fear. Many of his close kinsmen had died violently and had, in turn, exacted lethal revenge; he worried about being a potential target of retaliations and made it known to all that he would have nothing to do with raiding.[26]

Although human reciprocity and justice are undoubtedly far more developed than anything seen in other animals, there is more common ground than most students of law and ethics are aware of. Thus, reading *A Theory of Justice,* an influential book by the contemporary philosopher John Rawls, I cannot escape the feeling that rather than describing a human innovation, it elaborates on ancient themes, many of which are recognizable in our nearest relatives. Of course, everything is more explicit in human society because of our ability to formulate rules of conduct, discuss them among ourselves, and write about them in exquisite detail. Still, it is safe to assume that the actions of our ancestors were guided by gratitude, obligation, retribution, and indignation long before they developed enough language capacity for moral discourse.

A century ago Thomas Henry Huxley claimed that a purely competitive view of life cannot accommodate moral behavior.

> [The pursuit of virtue] repudiates the gladiatorial theory of existence. It demands that each man who enters into the enjoyment of the advantages of a polity shall be mindful of his debt to those who have laboriously constructed it; and shall take heed that no

act of his weakens the fabric in which he has been permitted to live.[27]

Instead of concluding that morality is a cultural construct that flies in the face of nature, Huxley and his followers would have done better to broaden their perspective on what the evolutionary process can accomplish. Petr Kropotkin had an inkling; Trivers formulated it with admirable precision. Given what we know now, the above statement about the advantages of mutual aid seems to hold reasonably well for monkeys, apes, and some other animals.

And could not the admonition at the end apply to Gwinnie?

5

GETTING ALONG

> Some stories cannot be literally true,
> and among these are both our current
> guiding myths—not only the Genesis
> story, but the Social Contract myth as
> well. The point is not just that there
> never was a moment of contract.
> Much more deeply, there was never
> the need to which that contract would
> have been an answer.
>
> *Mary Midgley*[1]

In the 1960s and early 1970s, the funding floodgates for research on aggression were thrown open by a single popular and hugely controversial book. Konrad Lorenz's *On Aggression* defined a new area of inquiry. Even his adversaries admitted as much, and took advantage of the research opportunities. Entire libraries started to be written on the topic, and soon a steady stream of articles compared definitions and explanations of the phenomenon. Though still a graduate student at the time, I was allowed to join a group of Dutch psychiatrists, criminologists, psychologists, and ethologists who regularly came together to discuss the roots of aggression and violence.

In this stimulating intellectual climate, I embarked on a study of aggressive behavior in long-tailed macaques—fairly small, brown-green monkeys native to Indonesia, Malaysia, and the Philippines. Working with a tape recorder and primitive black-and-white video equipment—equipment so heavy that setting it up was a back-breaking enterprise—I would patiently wait for spontaneous fights in a captive group. What struck me most while sitting and waiting was how rarely these monkeys fought, even though they had a reputation for belligerence: I calculated that they devoted less than 5 percent of their time to this activity. Most of the time they played, groomed, slept in large huddles, or were otherwise peacefully occupied. Then in a flash they would be chasing one another, aggressors barking and victims screaming in shrill voices, with the ever-present danger of serious damage from the males' sharp canines. I became as fast as a play-by-play sports announcer in speaking what happened into a microphone, recording in great detail the maneuvers of a number of simultaneous actors. The next moment the combatants would settle down, and I would be out of work again.

The common view of aggression as the expression of an internal drive, or a sign of frustration, or a response to some irritating external stimulus rapidly lost appeal for me during this period. Not that these notions were wrong—they were heavily debated, and there was something to be said for each of them—but they stressed the individual, not the social context. To me, the individuals became more and more of an abstraction; it was their relationships that caught my attention. In the same way that, when gazing at the night sky, we see the Great Bear or Orion instead of a mass of stars, I saw a dominance order, kinship bonds, alliances, and rivalries among my monkeys. Aggression was our label for the sparks that flew when interests collided: it could only be regarded as an interindividual, not an individual, phenomenon. It also was not a phenomenon that could be investigated in isolation, as it was so clearly embedded in other aspects of social life. How did the 5 percent and the 95 percent mesh? I did not solve the puzzle at the time, but sometimes felt I was studying the yang without the yin.

The questions emerging from this project, however vague in my mind, prepared me for the answer when it appeared a couple of years later during my chimpanzee studies at the Arnhem Zoo. If this sounds like the Holy Apparition Theory of Scientific Discovery, it is often how insights are achieved. Lorenz himself notes how "one day, after a long period of unconscious data accumulation, the *Gestalt* that has been sought is there, often coming completely unexpectedly and like a revelation, but full of the power to convince."[2]

The *Gestalt* that gave me the key to my problem was an embrace and kiss between two chimpanzees shortly after a serious altercation. The event caused hooting tumult in the colony; the other apes evidently attached great significance to what had happened. Since the embrace occurred between the chief opponents of the preceding fight, it struck me as a *reconciliation*. This insight was no doubt facilitated by the fact that chimpanzees are so humanlike in everything they do; we know now that reconciliations are not limited to these apes.

Since that day twenty years ago, I have watched literally thousands of reconciliations, highly emotional to rather casual, ones that appeared to resolve tensions to those that did not. It still gives me special pleasure to see the rapprochement phase (which may be hesitant and tentative), the physical reunion, and the calming talk or grooming that ensues. While this pleasure derives from my human capacity for empathy, I also have the reassuring knowledge that aggression has been put into context. Instead of the evanescent fireworks of group life, aggression turns out to be an integral part of social relationships: it arises within them, it upsets their dynamics, and its harmful effects can be "undone" through soothing contact.

Every reconciled conflict is a choice against entropy. As with expressions of sympathy, conflict resolution by peaceful means would never have come into existence were it not for strong attachments based on mutual dependency and cooperation. Resolution reflects a level of social connectedness lacking in many other species, such as the territorial birds and fish studied by early ethologists. Their choice of subjects explains why aggression was initially depicted as a *spacing* mechanism, a way of maintaining individual distance and defending territories. Observations of monkeys and apes challenge this notion: aggression followed by reconciliation is the exact opposite. Instead of dispersal, we see mutual attraction after fights. This pattern may seem counterintuitive, but not if one knows how strongly these animals rely on each other; they just cannot afford alienation.

To view aggressive behavior as an expression of interindividual conflict, both determined by and functional within social relationships, removes us ever further from the Lorenzian "instinctivistic" view; it pays equal attention to social and biological factors. Of course, a broad definition of the biological covers the social (social predispositions are as much part of a species' biology as its physiology and anatomy), yet the term is not commonly used this way in the eternal debate about aggression. A 1986 document known as the "Seville Statement on Violence," for example, tries to free humanity from what it calls "biological pessimism" by an outright rejection of genetic explanations. Oddly enough, no one would ever bother to write a similar manifesto to dispute the genetic basis for attachment, cooperation, or sex. Most people accept these behavioral universals as core elements of human nature, so why not aggression?

The reason is that aggression is the one trait that our species does not like to see when it looks in the mirror. It is the ugly pimple on our face, and biologists are blamed for suggesting that it might be hard to get rid of.

The Seville Statement targets outdated views; it still lives in a time in which we were being depicted as Killer Apes.[3] Contemporary ethologists include environmental factors along with genetics in their analyses of aggressive behavior, and have accumulated detailed data on the social context and resolution of interindividual conflict. While writing *Peacemaking among Primates,* I was astonished to discover that we know more about how our animal relatives cope with aggression than about how we ourselves do. Most research on human aggression still isolates the phenomenon from other aspects of social life, and treats it as an antisocial rather than a social activity.[4]

In this penultimate chapter, I provide an update of the growing body of knowledge about primate conflict resolution. I challenge social scientists to move from a paradigm defined in opposition to a thirty-year-old book to one that looks at human aggression from *both* a biological and a social perspective. This viewpoint can be achieved without in any way reducing aggression to an uncontrollable drive or instinct.

The Social Cage

Occidental culture enjoys a long-standing love affair with personal autonomy. We are individualists at heart, and what better way to stress this fact than by claiming that our primogenitor was self-sufficient? Jean-Jacques Rousseau engraved forever in our souls the image of the noble savage napping alone under the tree from which he had just collected his meal, satisfied in body and mind. What could he possibly need others for? His independence was not even compromised by love making: members of the two sexes simply forgot about each other once their passions had been stilled.

The total absence of social connectedness in this fantasy world did not bother Rousseau, who was a timid, retiring man most at ease during solitary botanical outings. For him, being *seul* was synonymous with being *libre.*

Though presented as a mere theoretical construct, not a historical truth, Rousseau's powerful image of autonomous savages lingers. It has become an article of faith among economists, who describe society as an aggregation of Robinson Crusoes, that is, a number of independent households engaged in free exchange. The same notion permeates theories of justice as a set of fairness principles agreed upon by free and equal human beings who have decided to compromise their autonomy so as to build a community. The concept of "rights" results: a morality that asks each person to respect the claims of each other person. John Rawls goes so far as to present the "initial situation" of human society as one involving rational but mutually disinterested parties.

Disinterested? As if our ancestors did not descend from animals that had lived for millions of years in hierarchically structured communities with strong mutual attachments. Any system of justice arising among them would have concerned individuals who were unfree, unequal, and probably more emotional than rational. There can be no doubt that they must have felt deep interest in one another as parents, offspring, siblings, mates, allies, friends, and guardians of the common cause. A morality exclusively concerned with individual rights tends to ignore the ties, needs, and interdependencies that have marked our existence from the very beginning. It is a cold morality that puts space between people, assigning each person to his or her own little corner of the universe.

How this caricature of a society arose in the minds of eminent thinkers is a mystery. Are we so painfully aware of the ancient patterns that we crave an antidote? Fairness and mutual respect do not come easily—we look at them as accomplishments, goals we fight for—and therefore, rather than admitting that we started out close but unequal, we like to give our quest for justice weight by constructing a story about how we were originally distant but equal. Like a nouveau riche claiming old money, we twist history to legitimize our vision of society.

The myth has served its purpose. With the French Revolution more than two centuries behind us, we now should be able to afford thought experiments that genuinely reconstruct the past. Although Rousseau himself briefly considered biology as a source of inspiration, he dropped the idea, lamenting the lack of knowledge in his day. Present chances to achieve a credible, scientifically grounded reconstruction are much better. We do not need to replace the noble savage with a killer ape, but we do need to abandon the romantic notion of an ancestor free as a bird.

While a wild male chimpanzee may roam a huge stretch of forest, there is a border beyond which he cannot venture, given the hostility between communities. He absolutely needs to get along with his male group mates: united they can stand against (and commit) brutal acts of territorial aggression. At the same time, he vies with these very same males for dominance. He must constantly keep track of his allies and rivals, as he may owe his rank to the first and run risks in presence of the second.

A chimpanzee female generally stays in a smaller home range within the males' territory; she is always surrounded by males who dominate her and offspring who appeal to her for assistance. I do not mean to imply a purely stressful existence, because wild chimpanzees also enjoy companionship, leisure time, and all sorts of gratifying activities—but I do believe that they are social captives.

It is like life in a small village where everyone knows everyone else: there is stability, closeness, and security in these relationships, but also intense control. This characteristic applies even more to monkeys than to chimpanzees. Chimpanzees live in so-called fission-fusion societies; that is, they move through the forest in small parties of a few individuals at a time, the composition of which changes regularly. All associations, except between mother and dependent offspring, are of a temporary nature.[5] Most monkey species form cohesive groups with all members present at all times; their social cage is even more confining.

Assuming that the face-to-face group with its attendant lack of liberty is the original condition of humankind, chimpanzees in captivity, such as the ones I work with, represent a unique experiment. Given the fluidity of their social network in the natural habitat, forcing them together in a permanent group seriously challenges their *adaptive potential,* defined as the range of conditions to which animals can adjust without compromising their health, ability to reproduce, or major portions of their species-typical repertoire of behavior. This range usually stretches well beyond the various habitats occupied in nature. The experience of the last two decades at enlightened zoos and research institutions, beginning with Arnhem Zoo's seminatural enclosure, demonstrates that chimpanzees are readily capable of living healthy lives, both physically and socially, in large captive colonies even if this situation implies forfeiting fission-fusion opportunities.

Under these conditions, social behavior resembles that of wild chimpanzees in many respects except, of course, that group life is considerably intensified. Adult females in particular show a social closeness unknown in their wild counterparts, who tend to be rather solitary. In captive settings, female chimpanzees protect each other against male violence, actively influence male power struggles through concerted action, groom one another regularly, and share food—behaviors rare or absent among females in the best-known wild populations.[6] Adult male relations seem less affected by living conditions; both in the wild and in captivity, males show a characteristic mixture of bonding and rivalry. In short, the chimpanzee's adaptive potential encompasses life in permanent associations similar to those of monkeys.

To say that these chimpanzees do not behave "naturally" misses the point. Few humans do. We spend long periods alone in small boxes on wheels, eat cooked foods, are in touch with relatives via wire, deal with strangers on a daily basis, and so on. Given this wholly artificial world, we may be pushing our adaptive potential to the limit; for this reason, the adaptability of other primates deserves close attention. People in modern societies and captive chimpanzees both successfully adjust to novel circumstances through psychological and social plasticity, which in and of itself is a natural ability.

The chimpanzees' capacity to live in stable groups is highly relevant in relation to human evolution. Without a similar capacity our ancestors could never have made the monumental step toward life in permanent settlements. As noted by Roger Masters, an American political scientist, the study of captive chimpanzee colonies may

> provide particularly valuable evidence of what happens to hominoids when they cannot easily leave their home community, and thus, like humans after the invention of agriculture, tend to substitute shifting social alignments for the group fission characteristic of hunter-gatherers and wild chimps.[7]

This observation brings us back to the issue of freedom. Regardless of circumstances, chimpanzees, monkeys, and humans cannot readily exit the group to which they belong. The double meaning of "belonging to" says it all: they are part of and possessed by the group. Migration between groups does occur in nature but tends to be limited to a particular life stage of one or the other sex. Despite opportunities for temporary distancing—opportunities that are considerably greater in the wild than in captivity, and greater in a populous "urban jungle" than in a village—in the long run, most group members are stuck with one another. My phrasing here is negative only to contrast the actual situation with the myth that we started out as a bunch of loners free to go wherever we wanted. Being social animals, we have no trouble recognizing the positive side. The definition of social animals is precisely that they seek and *enjoy* company; the social cage is their palace!

If group life is based on a social contract, it is drawn up and signed not by individual parties, but by Mother Nature. And she signs only if fitness increases through association with others, that is, if sociable individuals leave more progeny than do solitary individuals. We are seeing how social tendencies came into existence—via a genetic calculus rather than a rational choice. Even in our species, which prides itself on free will, we may find an occasional hermit who has opted for reclusion; yet we never encounter someone who has consciously decided to become social. One cannot decide to become what one already is.

It is not true that for us, or any other social species, life in groups has only pros and no cons; conflict resolution is necessary precisely because group life is not all upbeat. Its greatest disadvantage is that one is continuously surrounded by individuals searching for the same food and attracted to the same mates. Groups are breeding grounds for strife and competition, which ironically also pose the strongest threat to their existence. In an anonymous aggregation, animals may get away with back stabbing—in the way that schooling piranha occasionally take a quick bite out of each other—but an individualized society is doomed without checks and balances on aggression. In such a society, disruptive behavior has more consequences for the perpetrator as others tend to remember *who* is responsible. Cooperative relationships depend on trust that the other parties will act in a particular way under particular circumstances; troublemakers have difficulty gaining such trust.

I am assuming here that animals do not dampen competition for the sake of the group: the good of the group is a factor only insofar as it overlaps with that of the individual. Yet this overlap is considerable if group membership is a matter of life or death. Even apes have natural enemies, such as lions or leopards, and all monkeys suffer predation by birds of prey, snakes, and cats. The main benefit of group living, therefore, may be antipredator defense. Carel van Schaik and Maria van Noordwijk, two Dutch field biologists, traveled to different sites in Southeast Asia to see if group size in long-tailed macaques varied with predation pressure. One site, in Sumatra, had clouded leopards, golden cats, and tigers, whereas the other site, an island in the Indian Ocean, was cat free. Their study confirmed that in the absence of predation monkeys tend to live in smaller groups. In a theoretical paper van Schaik summed up: "The picture that emerges is quite simple: the lower threshold to group size is set by predation, the upper limit is posed by feeding competition among group members."[8]

Sue Boinsky, who followed squirrel monkeys in Costa Rica, noticed how females, upon giving birth (which they do with remarkable synchrony), shift from foraging on their own to traveling in close-knit groups. New mothers often sit side by side staring into the sky, continuously swiveling their heads. Boinsky ascribed their increased gregariousness to vigilance against birds of prey, and the fact that two pairs of eyes cover a wider angle than one.

> The amount of time that raptors were either directly above or near the troop rose from 2 percent to 10 percent when the babies were popping out like popcorn. Eight times I saw collared forest falcons try to grab a baby right off its mother's back; each time the falcons were mobbed and driven off. Some of the mothers that lost their babies had scratches and bleeding backs like those of monkeys I have seen attacked by falcons.[9]

The predator as the ultimate reason for group life is an idea with general applicability. Even elephants are thought to travel in herds because of the vulnerability of their calves to hyenas and lions. If this situation exists on land, imagine three-dimensional space, where surprises can come from any direction. In *Dolphin Days* Kenneth Norris discusses the features of dolphin "schools" that help them evade dangers lurking in the water. I use the quotation marks because an aggregation of dolphins is something quite different from a school of fish. Dolphins may look alike yet they recognize one another by special whistle sounds, and they lead complicated lives, including nursing of the young and lasting cooperative relationships, in a world in which they cannot lag behind for a minute. Even when dispersing in relaxed moments, they stay within what Norris calls their *magic envelope* of sights and sounds, a concept not unlike that of the social cage. When sharks or killer whales approach, the envelope tightens, sometimes in a flash, so as to confuse the predators with a coordinated response faster than their eyes can follow. It is virtually impossible to effectively pick out a potential victim among identically shaped, identically moving objects with little room in between.

Dolphins are a unique combination of a highly individualized society on the inside and faceless ranks on the outside. Norris describes their schooling reaction to Big Bear, a fisherman on a large vessel:

The dolphins leapt ahead of us, pouring from the water in ragged ranks. Suddenly Big Bear, steering with one hand, reached inside the cabin door for a rifle and began pumping shells in the water behind the dolphins to see, he said, if he could 'drive them.' The impact of those shells in the water near the dolphins must have been concussive, for they leapt out of the water as one, so tight to one another that you could not see between them.[10]

Apart from the survival value of belonging to a group with its multiple eyes and ears, its defense mechanisms, and (in case of social carnivores, such as wild dogs) its ability to overpower prey too fast or strong for a single predator—all of which concern external factors—there is the possibility of within-group collaboration to attain within-group goals. Many primates excel at this: their main tool for internal competition is the alliance of two parties against a third. Inasmuch as each individual tries to establish the best alliances, partners become commodities and social bonds become veritable investments, as Hans Kummer was the first to note.

Imagine that monkey A has managed over the years to cultivate a mutually supportive relationship with B, which allows the two of them jointly to keep lower echelons in line in the harsh manner typical of female macaques and baboons. A and B confirm and foster this partnership through mutual grooming and coordinated actions. Should A accept it when a third female, C, grooms B? If C is B's kin or of high rank, there is little A can do about it; blood is thicker than water, and the hierarchy is usually respected. But if C is neither, A will try to threaten C away from B, or simply push her aside. Such competition is commonplace and well documented in monkey groups indicating that social investments are carefully safeguarded.

Or take the triangle of Nikkie, the young alpha male of the Arnhem chimpanzee colony, his older ally, Yeroen, and their chief rival for the top spot, Luit. The extreme predictability with which Nikkie used to keep the other two males apart made me label it a "divide-and-rule" strategy. As soon as he saw Yeroen and Luit together, he would go over to break up the combination: not a single exception to this pattern was observed during Nikkie's years of leadership. Another indication of how much Nikkie valued his partnership with Yeroen—without whose support he could never have subdued Luit—was his eagerness to reconcile with him. Nikkie and Yeroen sometimes had tensions about mating rights. Initially Yeroen, the kingmaker, mated freely with the females; but over time the king himself became increasingly confident and bossy, hence less tolerant of his partner's sexual affairs.

Nikkie's disruptions of Yeroen's mating attempts invariably led to high drama, with Yeroen screaming and rallying female support and Nikkie nervously keeping an eye on Luit, who would grow in size and boldness as soon as the alliance was split. The only way for Nikkie to stay in control was quickly to mend the political base of his position, which he did by stretching out a hand to Yeroen. The ensuing embrace of the ruling males signaled the end of Luit's brief period of latitude. With Yeroen's backing, Nikkie would then make one of his spectacular jumps over Luit, who ducked for this display of dominance, shielding his head with both arms.

That Nikkie could not afford to estrange Yeroen became evident when the alliance finally collapsed after three years of cooperative rule. Overnight Luit took over.

The Relational Model

Constraints on competition, like that between Nikkie and Yeroen, are hard to accommodate within the received scientific view of aggression as an individual expression, a stimulus-response pattern, or a personality trait. Let us call it the *individual model*. It inspired classical studies such as the one in which rats attack each other after having been shocked by the experimenter, or in which human subjects are instructed to punish strangers or to report their feelings after having watched a violent movie.

Without denying its merits, I feel the individual model fails to give primacy to the most common context of aggression: relationships among familiar individuals. These relationships could, of course, be construed as complex stimulus-response chains, yet I favor a more direct focus on social context. The *relational model* views aggressive behavior as resulting from conflicts of interest between individuals who share a history (and a future). It assumes an equilibrium between tendencies that pull individuals apart and those that bring them together. It focuses on individuals drawn together by attachment and a sense of belonging to the same group. Whenever open conflict shakes this emotional foundation, it causes discomfort not unlike that associated with physical separation. Because conflicts may occur despite a conflux of interest between the same parties, aggression usually is not a one-time event but part of a series of encounters, from positive to negative, through which the relationship cycles.

In an attempt to explain why, in his words, "we should be so constructed that we find comfort in companionship and seek it, and that we experience greater or lesser degrees of anxiety when alone," John Bowlby, the British psychiatrist who founded attachment theory, suggested that avoidance of being alone ultimately derives from avoidance of danger.[11] As we have seen, the same can be said of conflict avoidance. Each clash with a valued companion brings in a waft of chilly air from outside the social cage.

To pursue this parallel between physical separation and emotional distancing due to conflict, consider the logical opposite of separation anxiety, which we may call *reunion euphoria*. After a separation, elephants greet each other with much spinning around, urinating, ear-flapping, entwining of trunks, clicking of tusks, and a chorus of rumbles and piercing trumpets. Cynthia Moss has no doubt that they experience joy: "It may not be similar to human joy or even comparable, but it is elephantine joy and it plays a very important part in their whole social system."[12]

The greeting ceremonies of chimpanzees, both in the wild and in captivity, are well known; they involve charging about with erect hair, loud hooting, kissing, and embracing. There is always a hostile undercurrent, which explains the gestures of submission and appeasement with which subordinates approach dominants. I once watched a reintroduction of two captive adult males who had lived together previously but had not seen each other for nearly five years. We were concerned about the potential for a fight (common between unfamiliar males), but they embraced and kissed in great excitement, slapping each other's backs as old friends, and groomed with tooth-clacking for a long while afterward. Tensions did emerge the next day, but the initial response was undeniably one of elation.

I have experienced human-ape reunions firsthand. Each time I return to the Arnhem Zoo, even if I appear among hundreds of visitors, the oldest female, Mama, will move her arthritic bones to meet me at the moat's edge with pant-grunts. Whereas Mama and I go back a long way, I have been similarly greeted by apes who I thought might have forgotten me.

Kevin, an adolescent male bonobo, had been transferred from the San Diego Zoo, where I had known him, to the Cincinnati Zoo, which I visited six years later. Let me add that, at least in observational studies, I make a point of interacting little with my subjects,

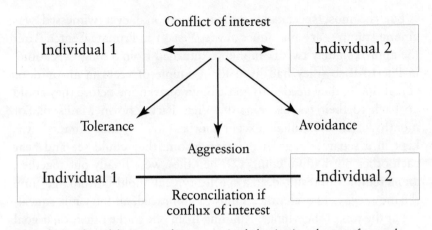

The relational model assumes that aggressive behavior is only one of several ways in which conflicts of interest can be resolved. Alternative solutions are the sharing of resources, priority claims based on previous encounters, or withdrawal from confrontation. If aggression does flare, the value of the relationship determines whether repair attempts will be made afterward; reconciliation is most likely if there is mutual interest in the maintenance of a valuable relationship. Parties negotiate the terms of their relationship by going through cycles of conflict and reconciliation.

and I certainly avoid feeding them or being directly involved in management. I want to reduce the attention they pay me; indeed, I feel honored if primates who jump up at the sight of a caretaker, veterinarian, or unfamiliar visitor accept my presence with hardly a glance, as if I have become part of the furniture. I do always say hello when I arrive, and with Kevin I recall a few close and friendly encounters in his night quarters.

Initially, Kevin just stared when I approached in the company of people familiar to him. As soon as I squatted down and said a few words, however, he pirouetted, clapped his hands, and invited me to play by locking eyes with me, then running away and looking over his shoulder. Distinctly happy to see me, he presented his back and started laughing with guttural sounds even before I dug my fingers into the base of his neck and under his arms (where all hominoids have their tickle spots). His immediate trust was so obvious that I was introduced later that day to my audience as "a good friend of Kevin," and most questions after my presentation were about our relationship. It also made me rethink my assumptions. I myself never forget a primate face that I have seen on a daily basis, and I certainly recognized Kevin instantly. Why had I expected it to be any different for him?[13]

But the most spectacular primate reunion I ever witnessed was among stump-tailed macaques at the Wisconsin Primate Center. After the approximately twenty monkeys returned from a long "vacation" in Puerto Rico (they had lived for a couple of years in an outdoor corral there), they needed to go through quarantine before they could go back to their enclosure at the Vilas Park Zoo in Madison. For months, each individual (except mothers with young infants) was kept in a separate cage in the same room: they could see and hear each other but had no contact. When they were finally put together, an incredible din resulted. In a species-typical "hold-bottom" gesture, the monkeys held each other's hips or embraced, all the while squealing at the top of their lungs. Running from one to the other, their goal seemed to be to go through this procedure with every possible partner, youngsters included. The excitement lasted for about half an hour, after which they settled down to groom each other with softer, more contented-sounding calls.

Given this celebration of togetherness, it is not surprising that stump-tails also reconcile fights more readily than almost any other species. A definite link exists between greeting and peacemaking: the first unifies parties who have been physically apart; the second, parties whose relationship has been strained. Both processes reflect the value attached to the relationship, and both confirm that, whatever has happened, the basic emotional connection is still intact.

Peacemaking

Golden monkeys do it with mutual hand-holding, chimpanzees with a kiss on the mouth, bonobos with sex, and tonkeana macaques with clasping and lipsmacking. Each species follows its own peacemaking protocol. Many have evolved gestures, facial expressions, and calls specifically for this purpose, whereas other species lack such behavior: their reconciliations look the same as any other contact.

How do we know, then, that these creatures are reconciling? Because the criterion is not a particular behavior but a particular sequence of events. I define reconciliation as a reunion between former opponents not long after their fight. A few acccidental reunions are not enough: reconciliation is believed to occur only if there is a systematic increase in friendly interactions following aggressive ones. This pattern is verified by comparing behavior after conflict with behavior at other, more relaxed moments. Quite a few studies have been made and have concluded that former opponents are *selectively attracted*, that is, they tend to come together more often than usual, and more often with each other than with individuals who had nothing to do with the fight. The phenomenon seems widespread in the primate order: we now have systematic evidence for chimpanzees, bonobos, mountain gorillas, golden monkeys, capuchin monkeys, red-fronted lemurs, patas monkeys, vervet monkeys, baboons, and a variety of macaques (rhesus, tonkeana, stump-tailed, long-tailed, pig-tailed, bonnet, Barbary, and Japanese macaques).[14]

Reconciliations range from the rather uneasy and nervous encounters of patas and rhesus monkeys to the seemingly pleasurable eroticism of bonobos. Some species reconcile only a fraction of conflicts, whereas other species reconcile a majority. Although this variability is as yet unexplained, it is logical to assume that the more conciliatory a species, the more important group cohesiveness must be in the wild. We assume that making peace is an evolved adaptation to the natural environment. Not that it is a relatively immutable trait: reconciliation is such a complex skill that it is probably strongly modified by social experience.

Primates begin developing this skill early on. As with everything related to attachment, it starts with the mother-infant bond, and it gets its strongest impetus from the inescapable weaning trauma. The mother pushes the infant away from her nipple, yet allows it to return right away when it screams in protest. The interval between rejection and acceptance lengthens with the infant's age, and conflicts turn into major scenes. Eventually, the infant learns to time its demands so as to reduce the likelihood of maternal rejection. Older infants have become quite smart at assessing maternal accessibility, as noticed by Jeanne Altmann in baboons in Kenya:

> Sometimes I was following a mother and could not locate her infant at all. The mother began a grooming interaction or resting and immediately the infant would dart toward her from an activity over 20 meters away, make contact, and begin suckling.[15]

Maintaining a tie with the mother despite occasional discord lays the groundwork for all later conflict resolution. Reconciliations with peers, probably next in importance, appear early in life as well. An early observation in our developmental study involved two four-month-old rhesus females, Oatly and Napkin. The two were play-wrestling when they were joined by Napkin's maternal aunt. This adult "helped" Napkin by holding her playmate down. Napkin took advantage by suddenly jumping on Oatly and biting her. After a brief struggle they broke up. The incident was not too serious, but it was remarkable for its aftermath. Oatly walked straight up to Napkin, sitting with the same aunt, and groomed her from behind. Napkin turned around and they embraced belly to belly. To complete this picture of harmony, the aunt then put her arms around the reconciling infants.

Unfortunately, observations of young monkeys tell us little about the learned nature of behavior. We can see a behavior become more complex with age, but in order to know how exactly it is acquired we need to manipulate experience. This goal became a major obsession with me after a debate with a child psychologist who, in defense of the virtual absence of data on human reconciliation, had argued that it was easier to collect such information on monkeys because they act in a stereotypical manner based on "instinct," whereas humans show perplexing variability. Apart from the fact that this could hardly be considered a valid excuse (the task of science is precisely to make sense of variability), differences between human and animal behavior are never that absolute. I decided that the field of primate conflict resolution could use a learning experiment.

My test paradigm combined two ingredients present at the Wisconsin Primate Center, where the project was conducted. One was the availability of two related primate species with contrasting temperaments. The other was the long-standing tradition of rearing experiments by Harry Harlow, the founder of the center. One of his students, Melinda Novak, had used so-called monkey therapists to rehabilitate rhesus monkeys who had been removed from their mothers and raised in isolation. These poor monkeys showed all sorts of stereotypies and tics as a result of their unnatural upbringing and were terrified of body contact. Novak's experiments, conducted in the early 1970s, demonstrated that they could be helped by bringing them in contact with normal members of their species. The isolated individuals learned from these therapists that contact can be comforting and desirable rather than frightening. Their stereotypies gradually disappeared.

Because my aim was not to see if I could cure abnormal monkeys but rather to see if I could change normal ones, I will speak of "tutors" instead of "therapists." We gave stump-tailed macaques an opportunity to act as tutors for rhesus monkeys. Stump-tails are easygoing, tolerant characters, whereas rhesus have a strictly enforced hierarchy. Most important, reconciliation occurs three times more often after fights among stump-tails than among rhesus monkeys, and stump-tails possess a much richer repertoire of reassurance gestures, including the hold-bottom contact that we first noticed during their grand reunion.

Our question was whether we could get any of the stump-tails' gentleness to rub off on the rhesus. We established mixed-species groups of two-year-old rhesus monkeys and two-and-a-half-year-old stump-tails (on the assumption that older, dominant tutors would be more effective than younger ones). These groups were kept together day and night for five consecutive months—a long time in the life of a macaque. Macaques reach adulthood at the age of four or five, so our cohousing experiment might be compared to placing a human child for two years in a chimpanzee colony (which, I wager, would have a profound, perhaps not altogether desirable, effect on the child).

When we first put the two species together, surprisingly, the rhesus were scared. Not only are stump-tails a slightly larger species, they are very tough beneath their gentle temperament; the rhesus must have sensed this fact. So, with the rhesus clinging in a fearful huddle to the ceiling of the room, the stump-tails calmly inspected their new environment. After a couple of minutes some rhesus, still in the same uncomfortable position, dared to threaten the stump-tails with harsh grunts. If it was a test, they were in for a surprise. Whereas a dominant rhesus would have answered the challenge and a subordinate would have fled, the stump-tails simply ignored it. They did not even look up. For the rhesus, this was perhaps their first experience with dominant companions who did not feel a need to forcibly underline their position.

In the course of the experiment the rhesus learned this lesson a thousand times over. Whereas mild aggression was common, physical violence and injuries were virtually absent; friendly contact and play soon became the dominant activities. At first the groups were strikingly segregated, reflecting the preference of almost any animal for its own species.[16] When Denise Johanowicz, the student who deserves much of the credit for the study's completion, arrived in the morning she would find a huddle of rhesus in one corner and a huddle of stump-tails in the other, suggesting that they had slept apart. Play between the species was unusual, but grooming did cross the divide. The stump-tails, as their name suggests, have hardly any tails themselves but were obsessed by those of the rhesus. We had the distinct impression that they sometimes groomed the rhesus just so that they could manipulate and inspect these strange appendages. Although the species segregation never totally disappeared, it diminished with time; toward the end of the experiment, the two species got along very nicely indeed, sleeping together in a single huddle.

Our most significant finding was that, after having lived with stump-tails, the rhesus reconciled more easily. Initially, they made up after fights as seldom as is typical of their species; but gradually they approached the high rate of their tutors, until they reconciled exactly as often as the stump-tails. Even after the stump-tails had been removed and the rhesus were left to interact among themselves, they maintained this newly acquired pacifism. Like chemists altering the properties of a solution, we had infused a group of monkeys of one species with the "social culture" of another.[17]

Imitation did not explain our results. The rhesus did not adopt any of the behavior patterns typical of stump-tails, such as hold-bottom gestures and teeth-chattering. Instead, they acted in every way like rhesus monkeys except for a distinctly friendlier disposition. The rhesus more often uttered a soft, pleasant-sounding vocalization, known as "girning," that signals good intentions during affiliation or play. Possibly the stump-tails' mellow, lenient temperament, combined with their unquestionable dominance, had created a social atmosphere that allowed the rhesus to loosen up and become more conciliatory than would be possible in a group ruled by the typical despots of their own kind. This explanation fits the relational model: different group dynamics had produced a fundamental change in the way conflict was being handled.

The outcome of this study contains an optimistic lesson. If rhesus monkeys are able to learn peacemaking skills, why not human children? Extrapolating to educational efforts in the human species, our findings suggest that the nature of relationships is the key. Reward or punishment of specific behavior by parents and teachers, such as the interruption of aggressive acts or the encouragement of peacemaking, may be too narrow an approach. Even if Western, particularly American, education will always emphasize self-actualization over group values, conflict resolution cannot be taught without attention to the social environment within which it functions. Perhaps we should look at techniques employed in other cultures. For example, Joseph Tobin and his coworkers compared preschools in different countries and observed that fights are almost never broken up by Japanese schoolteachers. These teachers follow the philosophy that children need to learn by themselves how to handle aggression and how to get along.

Evidently their education focuses on the general quality of relationships rather than on desirable or undesirable behavior.[18]

The most important factor by far is the *worth* of relationships. We are learning that primates, including humans, suppress aggression or make peace not for peace's sake but in order to preserve something valuable. This tendency was demonstrated by Marina Cords and Sylvie Thurnheer in an experiment with long-tailed macaques at Hans Kummer's laboratory in Zurich. In earlier tests Cords had provoked quarrels between two monkeys by giving a tidbit to one and not to the other. After the incident the monkeys tended to come together in a friendly manner. In order to create both reconciled and unreconciled pairs, the investigator sometimes prevented a reunion by distracting the monkeys. After this diversion the monkeys were presented with two drinking nipples, side by side, from which they could obtain a sweet drink. As expected, the reconciled monkeys drank together more readily than the unreconciled ones.

Cords and Thurnheer then took the bold step of increasing the worth of the relationship—a manipulation never tried before in primate studies. They trained certain monkeys to feed cooperatively. These monkeys were presented with a series of small holes, each just large enough for one individual at a time, through which they could reach popcorn, a favorite snack. In order to get the popcorn, they needed to take their positions at the same time, at holes closer together than their usual feeding distance and without exchanging any threats. The holes would remain shut if they deviated from these table manners. Once all the monkeys knew how to behave, they went through the same procedure as before, but instead of receiving drinks, they ended up at the popcorn dispenser. Since it was now impossible to get a reward without coordination with the other, the monkeys were smart enough to resolve the problem between them first: their reconciliation rate jumped more than threefold. Here we have the first firm evidence for *strategic reconciliation*, that is, reconciliation mindful of the partner's value.

Rather than a blind process, primate reconciliation is a learned social skill, sensitive to the social configuration of which individuals are part and wielded as an instrument to preserve precious ties. If this conclusion sounds rather calculating, we should not forget its emotional underpinnings. It is unlikely that primates rationally weigh how life would be with or without the other; they probably react to general anxieties about the relationship. If emotional closeness between individuals varies with their degree of mutual support, the rate of grooming between them, their kinship relation, and so on, then these feelings provide an excellent guide for "rational" decisions. A breakup or separation creating strong anxiety will almost always concern a relationship worth restoring. Several studies have confirmed that the tendency to reconcile increases with bond strength, as measured by the amount of time monkeys spend together.

The anxiety involved in the process has been studied by Filippo Aureli. Making use of the fact that monkeys scratch themselves at times of stress or insecurity—the way a student scratches her head at a difficult question—Aureli took careful measures of self-scratching after spontaneous altercations among long-tailed macaques. First, he established that the rate of scratching increases sharply after an individual has been attacked by another. Insecure about the relationship with the aggressor, recent victims are torn between approach and withdrawal. Their caution is well founded; often more attacks come their way, and not only from the initial aggressor.

Second, Aureli demonstrated that reconciliations, even brief ones, immediately reduce the scratching rate. If we accept that scratching reflects stress—which is supported by a variety of studies—friendly reunions between opponents alleviate stress. One reason is increased physical safety: conflicts followed by reconciliation are less often revived than those that go unreconciled. Further, reconciliation normalizes the relationship and restores tolerance and cooperation, as shown in Cords's experiments.

These results may sound blatantly obvious. What else would one expect? Yet all this needed to be proved before postconflict contact in nonhuman primates could justifiably be called reconciliation. Well, it happened of course the other way around, as so often in science—the label came first and much of the evidence later. Without studies such as those cited, use of the term "reconciliation" might have remained vulnerable to charges of anthropomorphism. Now that many predictions from the relational model have been tested and borne out by data, we can turn our attention to the model's full implications.

Rope Walking

Human aggression runs the entire spectrum from the little family quarrels, with which we are comfortable most of the time, to socially harmful forms, such as child abuse, rape, and murder. When the focus is on the most extreme expressions, it is too easy to generalize our concern and condemnation to all forms. The result is the unworkable mix of moral judgments, policy suggestions, and scientific insights that has plagued research in this field from the beginning.

To present all aggression as undesirable, even evil, is like calling all wild plants weeds: it is the perspective of the gardener, not the botanist or ecologist. Neither of the latter thinks in terms of the utility or beauty of plants, but studies their morphology, size, habitat, taxonomic status, and so on. From this perspective, the plant thrown out by the gardener is just as interesting and exciting, perhaps even more so, than the plant that is allowed to stay and flourish. It is this nonjudgmental, open-minded attitude that I advocate with regard to the question of how aggression functions in social relationships, and in society at large.

Even the verb "to function" is considered problematic because of its positive connotations. The "Seville Statement on Violence," for example, takes the interest of biologists in function to mean that they are out to "justify" aggression and violence, which of course they are not. Biologists rather make the point that a behavior that is universal in the human species, and widespread in the animal kingdom, simply cannot be as detrimental as social science would have us believe. There must be advantages associated with it—perhaps not for the recipients, but definitely for the performers. Biologists are not asking anyone to admire or encourage the behavior, only to step back enough to see that it is part and parcel of the social dynamics around us. So much so, in fact, that we may seriously question the tenet that aggressive behavior is by its very nature antisocial.

When demand exceeds supply—a common condition in nature as well as in human society—a collision of individual interests is inevitable. Some of the resulting conflicts will be resolved contentiously, perhaps with threat or use of violence. Therefore, we can dismiss absolute peace as utopian. Only two realistic alternatives exist in our imperfect world of limited resources: (1) unmitigated competition, or (2) a social order partly shaped and upheld by aggression.

Monkeys, apes, humans, and a host of other animals have clearly opted for the second possibility. Fighting fire with fire, we have made the use of force part of the solution to the problem of the use of force. One kind of aggression restrains other kinds, including the raw contest springing from unevenly distributed resources. Threats and intimidations signal interest in food and mates, back up or stake out priority rights (which in the long run help reduce conflict), or prevent physical harm to the young and weak by putting a halt to bullying by others. Through these regulatory functions an order is created that in complexity far exceeds that of animals, such as grazing herds of bovids, characterized by low levels of competition owing to evenly distributed resources.

Thus, some of humanity's most cherished social institutions are firmly rooted in and upheld by aggression. Systems of justice, for example, can be regarded as the successful transformation of a deep-seated urge for revenge—euphemized as retribution—which keeps this urge within acceptable boundaries. Law enforcement, which is little else than governmental violence—euphemized as force—is often, but by no means always, sanctioned by the majority of the population. There is also abundant historic evidence of violence as a means to bring about much-needed societal change, the threat of which makes elites realize that they can ignore the plight of the poor and the disadvantaged only at their own peril.

Notwithstanding the claim of the British ethologist Robert Hinde that the cumulative unhappiness produced by individual acts of aggression rivals the threat of the nuclear bomb, the above examples make it hard to deny the existence of constructive forms of aggression at the societal level. The same applies at the interpersonal level. Whether aggression has healthy or unhealthy consequences for social relationships depends on how and when it is used, how far it is allowed to escalate, and what is at stake. It also depends on the power balance: if aggression always comes from one side, and only serves the interests of one party, it obviously is not much of a constructive mechanism for the other party. It is a matter of dosage and context. In the same way that one can drink beer without being an alcoholic, one can express anger without being abusive.

We can all distinguish a mean-spirited attack from charges that express love and concern, such as the shouting and screaming of a parent at a child who has almost fallen off the roof. Anger is so much part of our daily relationships that it is rarely pure: it mixes with all sorts of other feelings. One of the most notable fictional sagas of "prosocial" aggression, at least from the perspective of the recipient, is cartoonist George Herriman's creation *Krazy Kat*. This feline is so much in love with Ignatz, a mouse, that every brick thrown by the mouse is perceived as a sign of affection. Thus, instead of stars we see little hearts around the cat's head when it is hit by Ignatz's bricks. And if no brick has been forthcoming for a while, Krazy Kat gets bored and says such things as "He'll not fail me—that dollin."

Recent data on human marital relationships have begun to chal-

Even if Krazy Kat and Ignatz have a somewhat dysfunctional relationship, brick throwing is not necessarily antisocial. (Cartoon by George Herriman; reprinted with special permission of King Features Syndicate)

lenge the conventional wisdom that the best way of handling conflict is invariably by "talking things out." Marriage counselors used to steer couples away from bickering and explosive fights, convinced that such conflicts poison the relationship. We prefer couples to fight with words rather than fists or heavy objects, yet one can hardly expect them to negotiate the most important relationship of their lives without occasional anger and frustration. Extensive research on marriage and divorce by the American psychologist John Gottman suggests that much more important than *whether* people fight is *how* they do it.

Gottman distinguishes three kinds of marriages: the *avoidants* (who dodge and minimize conflict), the *validators* (who carefully listen to each other's arguments), and the *volatile* (who argue and fight on a grand scale). Even though the last kind of marriage may seem to court disaster, the investigator observes that

> it turns out that these couples' volcanic arguments are just a small part of an otherwise warm and loving marriage. The passion and relish with which they fight seems to fuel their positive interactions even more. Not only do they express more anger but they laugh and are more affectionate than the average validating couple. These couples certainly do not find making up hard to

do—they are masters at it. As intense as their battles may be, their good times are that much better.[19]

I sometimes imagine people as having ropes between them. The thicker the rope, the harder they can pull each other around without breaking the tie. If it is less of a problem to express hard feelings to a spouse than to a colleague, it is because the thickest ropes best endure conflict. Thus, conflict engagement may follow a simple law: the more easily reconciliation can be achieved, the less reluctant we are to fire a salvo.

Lest this leave the impression that conflict and reconciliation are desirable under any circumstances, I should add that the rope metaphor comes with other images, including that of getting entangled. Sometimes a tie that is not worth preserving is kept alive by giving in too readily to the demands of the other—or by accepting apologies for unacceptable behavior. If aggression reaches damaging levels or individual interests are not served equitably, reconciliation becomes a maladaptive process. Relationships can unravel either because a tie is too weak to sustain serious negotiation or because one of the parties systematically shortchanges his or her own interests.

The relational model is agnostic about whether aggression, or reconciliation, is good or bad. Its main purpose is to study how conflicts of interest are worked out, under which conditions this process results in overt hostilities, and how these are dealt with. Aggressive behavior is viewed as one of a multitude of tools available to shape a relationship. Use of the tool may be compatible with harmonious, mutually beneficial relationships, may create tensions and depressions, or may result in the "traumatic bonding" sometimes seen between victims and victimizers. The exact interplay of factors makes us decide whether to resort to aggression or not, whether to emphasize our own interests or not, and whether to preserve a tie or not. I would argue that understanding the normal, acceptable forms of aggression needs to be part of any endeavor to understand the more severe forms that worry us.

Baboon Testimony

The view of social relationships as arenas in which interests are traded back and forth, partly via stormy encounters, recently received a boost from macaque research. It has been known since the pioneering fieldwork of Japanese primatologists in the 1950s that the strongest ties in macaque society are those of female kin. Grandmothers, mothers, daughters, and sisters often groom each other, tolerate each other around food and water, and band together against other matrilines. Now studies on both free-ranging and captive monkeys have demonstrated that the highest level of fighting occurs *within* the matrilines. Thus, rather than automatically marking poor relationships, aggressive clashes are common between individuals who by any other standard have strong and positive relationships.[20]

This finding should not be pushed too far. Related monkeys do not spend all their time squabbling—peaceful association is the rule. The point is that fights occur in the best of families, and that strong relationships are not necessarily characterized by the absence of conflict. As in the "volatile" marriages, what matters more than the frequency of conflict is how the conflict is weathered. If aggressive behavior serves to negotiate the terms of relationships, a high rate among individuals who are close and mutually dependent makes perfect sense. There is plenty of opportunity in long-term relationships to go through repeated conflicts so as to achieve a fine-tuning of expectations about each other. Once the terms have been accepted, the intensity of conflicts will diminish.

Every day I can see endearing compromise in the chimpanzees at the Yerkes Field Station. Some of the juveniles suck on their mother's lower lip or ear, or nap while sucking on the skin right next to her nipple rather than holding the nipple itself in their mouth—which they are no longer allowed to do. Such substitute nursing is the result of an extended period of discrepant desires between their mother and themselves, in which access to milk has become increasingly restricted despite vociferous protest. Toward the end of the process, youngsters often bury their face close to their mother's breast, in her armpit for example, only to sneak up to the real thing when they get a chance. Occasionally she gives in, but at other times a major falling-out results. Most four-year-old juveniles have learned not even to try the underhanded maneuver anymore; they may stick a finger or toe in their mouth, but they prefer to create an *illusion* of being nursed. We observers have to watch carefully, because from a distance one would swear it is for real.

Mother and offspring bring different weapons to the weaning battlefield. The mother has superior strength; the offspring, a well-developed larynx and subtle blackmailing tactics. After all, the mother has no desire to lose the time and energy she has put into gestation, nursing, and protection. The youngster cajoles her with signs of distress, such as pouts and whimpers, and—if all else fails—a temper tantrum, at the peak of which he may almost choke in his screams or vomit at her feet. This is the ultimate threat: a literal waste of maternal investment.[21] In addition, the commotion may attract unwanted attention to the mother, whereby an adult male, for instance, may hit her for causing all the noise. Thus, apart from undermining his mother's determination by raising concern about his well-being, the offspring exerts social pressure.

Wherever there is a clash of wills, negotiation takes place. Each party develops expectations about the other's behavior, which in time converge on a tacit agreement about what is acceptable. The agreement is not a rational decision, but an adjustment to how the other reacts to one's own behavior. Since the end product is that the wants and needs of the other are taken into account, we are looking at the prototype of a social contract.

The feelings guiding this adjustment are central to human morality, ranging from sympathy and a desire to please to anger and insistence on what one has come to expect. William Charlesworth, a child psychologist and ethologist, sees a child's protest against diminishing maternal care as the first expression of moral outrage.

> One of the infant's earliest perceptions of its world is in terms of how advantageous or disadvantageous it is to its well-being. What is advantageous is perceived as just; what is disadvantageous as unjust. Anger is . . . a clear-cut instance of the earliest sense of a disadvantageous or unjust world.[22]

Charlesworth is not saying here that the child is interested in justice for anyone except itself; that would require more advanced moral understanding than the child has. The negotiations I am describing are driven by self-interest, and just outcomes are by no means guaranteed.

Mother and offspring have so many interests in common that, with increased juvenile independence and the mother's readiness for the next pregnancy, the weaning conflict usually works itself out to their mutual advantage. Exceptions do occur, however. Jane Goodall watched Flo, who used to be an excellent mother, lose the battle as she became too old. Lacking the energy to put up with her last son's aggressive tantrums, Flo carried him on her back until she could no longer bear his weight: by that time he was eight. We also know how in our own species, if one party brings superior muscle power to the domestic arena and is willing to use it, equitable outcomes may as well be forgotten.

How conflicts of interest are settled depends substantially on the kind of relationship two individuals have, which in turn depends on their species. It is easy to see how the mother-offspring relation shares fundamental characteristics across all mammals, but other relations vary tremendously. Some species enforce dominance rigidly, thereby delineating the amount of negotiation space. As a result, members of different species often end up with strikingly different expectations about how they will and should be treated. Each species has its own sense of social regularity.

How much expectations matter became evident when I began working with capuchin monkeys. In a major departure from my previous hands-off approach, in the capuchin project I wanted to investigate social behavior experimentally. Rather than putting the monkeys in single cages to facilitate testing—as is still often done—I dreamed of keeping the entire colony together in a large pen with outdoor space, and the testing area in the same facility, so that subjects could stay vocally in touch with their group during the brief tests I planned. This arrangement seems to me the best way of maintaining captive primates. Moreover, it is the only way to work with monkeys who have long-term social relationships that they care about, which is essential for my research.

The entire plan hinged on our ability to separate individuals from the group without causing too much stress. To this end, they needed to learn to enter a carrier. We interacted freely with them so that they became used to us, and soon some of the capuchins were so tame that they would jump on my shoulder, or check my pockets without permission. A few of the older individuals, however, remained so shy that they would not even accept a peanut from my hand. I naively assumed that my "friends" in the colony would be the first to learn what I wanted, and that the timid individuals would take longer. Was I wrong! The shy ones quickly grasped that when I clapped my hands and chased them, their problems were over as soon as they jumped into the carrying cage. I then immediately released them into another pen with lots of good food; after a few trials they were more than happy to follow the procedure.

The tamest monkeys were the ones who gave me the most trouble. It was not just that they refused to go into the carrier—they were obviously less fearful of me—but they started a major brawl with this man whom they had come to trust and who had the temerity to violate all their expectations. They screamed and barked, lunged at me, slapped at my face, but also made friendly sounds in between their bouts of rage, as if pleading with me to call off the whole business. We had, after all, established a bond, and they very reasonably assumed that this gave them some leverage. Not wanting to hurt them, but also not prepared to let them get away with their behavior, I simply showed more patience and determination than they were willing to invest. In a few cases it took an hour before the monkey finally entered the carrying cage, leaving both of us exhausted. The same individuals would enter in ten minutes on the next day, then in five, until they would try to go into the transport cage even before I had opened it.

I still work with these monkeys today, and they must now be among the easiest in the world to handle. The Yerkes Primate Center made the setup possible, and the capuchins, smart and small, turned out to be an ideal choice. The strenuous disagreement the capuchins and I had to go through before reaching this point was entirely in line with the temperament of the species. Capuchin society is closer to the chimpanzee's tolerant arrangement than to the rhesus monkey's despotism.

This experience also made me appreciate the claim of Vicki Hearne, an American horse trainer, that trainers and their charges enter what she calls a morally loaded pact; that is, they develop expectations about each other through a process of mutual agreement, not plain coercion. With animals as physically intimidating as horses, it is perhaps easier to see why. Discussing her approach to a "crazy" horse named Drummer Girl, Hearne explains:

> If, for example, I were to approach her believing some febrile nonsense of the sort I quoted earlier, about training as coercion, then my beliefs would be manifest in my body as I approached her, and she, reading this, would plain old kill me, that's all, if she could find a way, thus underlining a remark of William Steinkraus's—that the horse is the ultimate authority on the correctness of our theories and methods. It is wrong to talk about coercion not only because that doesn't happen to be what's going on in places like the Spanish Riding School, where horses are trained to the highest pitch, but also because it cannot go on with horses performing at high levels: horses have souls, and there is an inexorable logic consequent on that which makes coercing them into high performance impossible. I don't mean that you

can't coerce horses—only that if you do you will end up, if you are lucky, with a dull, unenthusiastic mount or, if you are unlucky, with a Drummer Girl emerging murderously from the trailer.[23]

Although coercion and the threat of force are very much part of primate (including human) relationships, the ropes tying individuals together would fray quickly indeed if these tendencies were not held in check. A mother chimpanzee could stop her offspring's suckling attempts at once, and a dominant chimpanzee could keep all food for himself—but this is not how things are done if there is an interest in maintaining good relationships. So, rather than coercion, much of the time we see a process closer to *persuasion*.

Perhaps the best-known form of negotiation among primates is the way adult male baboons greet other males in their troop.[24] As described by Barbara Smuts and John Watanabe, who studied greetings in the field in Kenya, one male will typically walk up to another with a rapid, swinging gait. He looks the other straight in the eye with some friendly expression, such as lipsmacking, which makes it absolutely clear that he only wants to initiate a greeting. To communicate intention is essential, given the fierce rivalry of males over females and the formidable canines with which males can cause a deep gash in a split second.

The encounter itself follows a certain protocol that varies with the kind of relationship the two males have. Often, the other welcomes the approach with a similar friendly expression, and one male presents his rear end while the other touches or grasps his hips. They may then proceed to mounting, or, if they really get intimate, one male may fondle the other's scrotum or pull at his penis. Known as "diddling," it is a sign of tremendous trust. The contact lasts only a few seconds, after which the two males separate again. Male baboons do not seem comfortable enough in each other's presence to associate or groom; their predominant modes of interaction are fighting and greeting.

The same behavior was studied by Fernando Colmenares in a large colony at the Madrid Zoo. He found the encounters to be extremely tense, occasionally erupting into fights. The reason is that they often serve to test and confirm who is on top, hence the jockeying for position to decide which male is going to be the mounter (generally the dominant) and which the mountee. Greetings thus seem a way of assessing intention: a male who used to elicit presentation in another learns from his partner's refusal that their previous roles are no longer working, and that a serious challenge may be in the air. Since tensions remain under control in the vast majority of greetings, the advantage of this sort of information exchange is that matters can often be worked out without physical confrontation.

Smuts and Watanabe found the youngest, most pugnacious adult males to be the most preoccupied with dominance. They are extremely edgy during greetings. Senior males who usually have known each other a long time, are less nervous, and instead of competing have learned to cooperate. With one another, that is: their old-boy network effectively keeps younger and stronger males from claiming every attractive female. Concomitantly, the older males' greetings are conducted at a leisurely pace, as they have changed from negotiations about dominance to confirmations of partnership. A few combinations of older males have even reached the point at which dominance hardly matters anymore; they have moved from asymmetrical to symmetrical greetings.

Take Alexander and Boz, two devoted allies. On one occasion Boz heard Alexander scream from 50 meters away, hurried over to the spot, and jumped without hesitation on the back of his buddy's attacker. Alexander would have done the same for Boz. This alliance served them so well that every day, first thing in the morning, the two males would go through a series of intimate greetings so carefully balanced that one would think they were keeping count. Boz would present to Alexander to let him touch his genitals, while both gazed into the other's face and lipsmacked. Two minutes later Alexander would present to Boz for the reverse procedure. If they were agreeing on something, the symmetry of their encounters made it evident that it was on mutual support and shared profits for the fresh day, not on who could push around whom.

The formalization of roles and the remarkable involvement of vulnerable body parts made Smuts and Watanabe draw a parallel with biblical oaths in which one man places a hand under another's loins. Considering that the words "testify," "testimony," and "testicle" share a common Latin root, it does not seem too far-fetched for these primatologists to speculate that

> the genital touching that sometimes occurs in greetings perhaps serves to enhance the truth value of whatever these males are 'saying' to each other within the formally circumscribed context

of greeting. Lacking articulate speech, and unable to swear oaths, perhaps male baboons make a gestural equivalent by literally placing their future reproductive success in the trust of another male. Such risky gestures may help to enhance whatever verity is presumed in the greeting because they impose a potential cost on the presenting male.[25]

Draining the Behavioral Sink

Some of the most destructive aggression in modern society takes place between strangers or individuals who share few or no interests: it is the bloodshed of gang warfare, drive-by shootings, rioting, muggings, and random violence. This sort of aggression hardly fits the relational model, for it concerns people without ties; attackers and victims may not even view themselves as part of the same community. Much of this violence takes place in cities, and unfortunately (or perhaps fortunately) these massive collections of strangers have no parallel in the social life of monkeys and apes. Not that this fact has prevented the highest levels of those in the U.S. science and government from embroiling themselves in primatological discourse about the issue.

The matter burst into the open when Frederick Goodwin, the head of a large federal organization (the Alcohol, Drug Abuse, and Mental Health Administration), proposed at a committee meeting of the National Institutes of Mental Health on February 11, 1992, that monkey behavior may hold clues for an understanding of inner-city crime:

> If you look, for example, at male monkeys, especially in the wild, roughly half of them survive to adulthood. The other half die by violence. That is the natural way of it for males, to knock each other off. . . . Now, one could say that if some of the loss of social structure in this society, and particularly within the high-impact inner-city areas, has removed some of the civilizing evolutionary things that we have built up and that maybe it isn't just a careless use of the word when people call certain areas of inner cities jungles. (excerpted verbatim from a transcript of the meeting)[26]

These remarks caused such an outcry that Goodwin ended up resigning his position. Part of the problem was that a comparison between inner-city youth and male monkeys struck some politicians as racial stereotyping, even though Goodwin himself never mentioned race. Actually, the affair went much deeper, reflecting profound disagreement about the role of biology in human behavior.

On the one side stand scientists who believe that human aggression cannot be purely cultural. Genes, brains, and hormones must be involved, otherwise why would, for example, the vast majority of violent crimes all over the world be committed by young adult males? The other side is deeply suspicious of the policy implications they fear may emerge from a biomedical approach, such as the screening of children for possible genetic or neurological markers of a violent predisposition, and the decision on how to treat such children. It is the familiar problem of whether we want to gain knowledge that might lend itself to abuse. According to this school of thought, all we really want and need to learn about aggression is its socioeconomic and environmental dimensions.

In a peculiar twist, the chief spokesperson for this group, Peter Breggin, has argued that if comparisons with animals are going to be made at all, they should be with chimpanzees, not monkeys. In a letter to the *New York Times* (March 15, 1992), the psychiatrist explains: "The chimps are relatively nonviolent, family-oriented and settle conflicts within the community through loving activities, such as playing, hugging, kissing and mutual grooming. We would do well to mimic them."

I could almost have said that myself, and I agree wholeheartedly that we could learn a thing or two from these apes. Yet I also believe that the point can be made just as effectively without downplaying the lethal "warfare" observed between chimpanzees in the wild, and their occasional infanticide and cannibalism. Chimpanzees are no angels of peace. It is precisely because they *can* throw overboard all inhibitions that their usual restraint is so impressive.

With scientists publicly hitting each other over the head with primate examples, a golden era of research in this field seemed to be dawning until two politicians, Senator Edward Kennedy and Congressman John Dingell, issued the verdict that "primate research is a preposterous basis for discussing the crime and violence that plague our country today."[27]

Looming behind this entire controversy is the dreadful image of the *behavioral sink*. Intentional or not, Goodwin's remarks about the loss of civilization and structure in the inner-city "jungle" echoed the popular view of disorder and depravity under crowded conditions. We owe this perspective to John Calhoun's experiments on domesticated rats, published in 1962 under the title "Population Density and Social Pathology." Calhoun opened his paper with Thomas Malthus' observation that population growth is automatically slowed by increased vice and misery. The scientist added that while we know overpopulation causes misery such as disease and food shortages, we know virtually nothing about its impact on vice.

This reflection inspired the nightmarish experiment of an expanding rat population in a crammed room killing, sexually assaulting, and eventually devouring each other. Much of this scenario took place among the occupants of a central feeding area. Despite the presence of food hoppers elsewhere in the room, the rats were irresistibly attracted by the social turmoil of this particular area; as a result, great numbers of them jostled and trampled each other for a position at the hoppers. It was this magnetism of the crowd and the attendant chaos and deviancy that made Calhoun speak of a "behavioral sink."

In no time, popularizers were comparing politically motivated street riots to rat packs, inner cities to behavioral sinks, and metropolitan areas to zoos. Warning that society was heading for either anarchy or dictatorship, Robert Ardrey, an American science writer, remarked on the voluntary nature of crowding: "Just as Calhoun's rats freely chose to eat in the middle pens, we freely enter the city."[28] In jumping from rodents to people, however, these writers missed something terribly important: the connection between population density and aggression is far from straightforward in our species.

I am particularly skeptical myself, having moved from the country with the highest population density in the industrialized world to a country with lots of empty space and by far the highest murder rate: the United States has fourteen times fewer people per square kilometer than the Netherlands, yet ten times more homicides per capita. It is not simply a matter of whether people live in large or small agglomerations. Cities are the most crowded places, but why should Washington have an annual homicide rate of 34.6 per 100,000 inhabitants and New York 14.4, compared to Berlin's 1.4, Rome's 1.2, and Tokyo's 0.5? Population density cannot be the answer.[29]

What is most amazing is that our species is able to survive in cities at all, and how relatively *rare* violence is—as eloquently expressed in 1976 by the American ethologist Nicholas Thompson:

My students were New York citizens who rode the subways and daily saw the spectacle of hundreds of people of every age, sex, race, religion, mode of dress, smoking habit, eating habit, and degree of wakefulness compressed peaceably into an intimacy rarely shared between husband and wife in our society, much less by strangers. Yet despite this spectacle, my students were eager to interpret the handful of aggressive incidents which occur daily in the subways as evidence that man's simian nature ill suits him to live in an urban environment. To an ethologist, what is surprising about people in subways is not their hostility; on the contrary, it is the degree of coordination and habituation which permits thousands of people to move daily through an environment so physically hostile as to stampede the herds of any sane animal.[30]

Like Thompson's students, however, the general public preferred to listen to authors who exaggerated the problems of the city, to make the point that humanity lacks the capacity to handle the situation, and blamed this on our brutish past.[31] Initially, primatologists went along with this pessimistic position, claiming that captivity gives rise to bullying monkey bosses who terrify the group, that dominance hierarchies are mere artifacts of stress, and that the rate of aggression skyrockets when monkeys lack space. Combined with the then-prevailing view of total peacefulness in nature, scientists saw striking similarities between primates and rodents.

Things changed when field-workers began to report sporadic but deadly violence in a wide range of species—from langurs to gorillas—as well as strict and clear-cut hierarchies that remained stable for decades. They also measured remarkable hardship in low-ranking monkeys. For example, Wolfgang Dittus saw food competition result in what he termed "socially imposed mortality": no less than 90 percent of the males and 85 percent of the females born in a wild population of toque macaques in Sri Lanka died before reaching adulthood, mostly from malnutrition. Compared to the food abundance and absence of predators in captivity, one would think that stress would be higher, instead of lower, in natural habitats with such dim survival prospects.

At the same time that the Rousseauian image of primates in their natural state was crumbling, questions were being raised about the supposed connection between population density and aggression. It was felt that density needed to be clearly distinguished from other factors. For example, in 1971 Bruce Alexander and E. Roth reported serious fighting, even killing, when a group of Japanese macaques at the Oregon Regional Primate Research Center were released into a corral seventy-three times *larger* than their original enclosure. After then living for years in this corral, crowding the group into a small pen produced a similar rise in aggression. The investigators concluded that "removal from a familiar habitat is adequate to provoke mobbing and this effect is independent of the characteristics of the new habitat."[32]

As a consequence of these and other puzzling findings, new research tried to circumvent novelty effects of environments and companions through a focus on well-established groups in familiar habitats. The results were inconclusive, with some investigators reporting more aggression under crowded conditions, and others equal or even less aggression.[33]

Whereas it is logical to expect social friction under conditions of high density, we know that primates have ways of managing this problem. To see if their control techniques might be the reason why they respond differently than rats, one of my students, Kees Nieuwenhuijsen, carefully compared long-term behavioral records on the Arnhem chimpanzee colony collected over hundreds of hours of observation both indoors and outdoors. The indoor hall in which the colony spends the cold winters is only 5 percent of the size of the outdoor island, so we were dealing with an exceptionally high crowding factor. The apes were thoroughly familiar with both environments, which eliminated novelty effects.

In the winter the chimpanzees seemed irritable, at times tense, but not overly aggressive. For example, two females who normally would stay out of each other's way because they do not get along, had trouble doing so in the hall. A problem might ensue if the infant of the one naively crawled into the lap of the other. Rather than open conflict, however, this situation usually was resolved by the second female, even if she was dominant, moving out of the youngster's range. Alternatively, the mother might go over to her for a grooming session that would have been unthinkable on the island.

Males ready to challenge the existing order would "lie low" during the winters, frequently paying their respects to the alpha male by greeting him with bowing and pant-grunting. Opportunities to maneuver without being cornered, and to engage in one-on-one battles without all the others on the sidelines, would only open up after release onto the island. Within weeks trouble might start; almost all major power struggles in the Arnhem colony's history have occurred outdoors.

While the frequency of aggressive incidents went up less than two-fold in the winter hall compared to the island, their intensity barely changed, and we did not notice more injuries. The effect was very modest, given the crowding factor. Other kinds of behavior, such as submissive greetings and grooming, increased even more than aggression. Inasmuch as grooming and greeting serve appeasing and calming functions, they may have smoothed strained relationships. Most important, we found that some measures were affected during the first couple of weeks only. Their subsequent return to summer levels implied that the traditional test paradigm, in which animals or human volunteers are crowded for a brief period, measures initial reactions only, not any long-term effects of the situation.[34]

Because nothing remotely resembling the extreme violence and hypersexuality of the behavioral sink occurred with crowded chimpanzees, I developed an alternative theory. It is not a complete alternative, as it incorporates Calhoun's idea that crowding puts individuals and their social system under stress. The big difference is that the new hypothesis assumes that primates do not take this tension lying down: they utilize countermeasures to keep their society from collapsing. According to the theory, we expect the strongest crowding effects in "naive" populations facing this challenge for the very first time, and weaker effects in populations that have had time to adjust.

Preliminary support came from studies by Michael McGuire and coworkers on vervet monkeys, and from my own inspection of published reports on rhesus monkeys. Both comparisons concerned existing populations under free-ranging as well as captive conditions, and both indicated fairly uniform levels of aggression. Furthermore, crowding in a highly social nonprimate, the dwarf mongoose, was studied by Anne Rasa. She found a rise in aggression but also in affiliative behavior, resulting in stabilization of the social system. Possibly, then, coping mechanisms are not unique to primates.

The next logical step was to cast the net as widely as possible, and have a single investigator with a standardized method evaluate populations under various conditions. I selected groups of rhesus monkeys in small indoor pens and large outdoor enclosures at the Wisconsin Primate Center, in spacious outdoor corrals at the Yerkes Field Station, and monkeys set free on a large island off the coast of South Carolina. The four conditions differed dramatically in the number of monkeys per square meter: the island offered 646 times more surface area per individual than the small pens. And this is not even a fair comparison, as the island is covered with oak and palm trees: taking three-dimensional space into account, the crowding factor exceeded 6,000!

Peter Judge, an American primatologist, joined my team in 1989, as a traveling observer. Over the years he collected thousands of "focal follows" (observations on one monkey at a time) at the various sites, entering each one into a computer. Here is what we found.

Minor aggression increase: The average rate of aggressive acts per adult per hour ranged between 1.6 and 2.6 across conditions. Although the top rate was not reached under the most crowded condition, the general trend was for more aggression under higher densities. In view of the great variation in space, however, the effect on aggression was astonishingly small.

Increased aggression intensity: Under all conditions, most aggressive encounters consisted of mere threats and chases. Under crowded conditions, however, severe aggression, such as biting, increased more than the low-intensity forms and monkeys suffered more injuries as a result.

Sex difference: Only females showed the two trends above; male aggression and wounding were unaffected by spatial conditions. Perhaps the competition and the posturing that males must engage in to be successful is always high or at an upper limit. Females generally perform less aggression than males and often can avoid direct competition under low-density conditions. Unlike males, they have room for increase: their aggression level rises as they become more crowded.

Increase in grooming and appeasement: The hourly rate of grooming per adult ranged from 2.6 to 4.2 across conditions, increasing steadily in both sexes with population density. Submission and appeasement gestures also went up under crowded conditions.

These data suggest that if a monkey population lives long enough in a particular environment, aggression rates will stabilize at a species-typical level that varies little with density. The social system, with its matrilineal hierarchy and kinship network, remains essentially the same under all conditions, and apparently gives rise to similar amounts of social friction. In addition, monkeys groom each other more and show more appeasement gestures when densities are higher, which may dampen tensions. The one detrimental correlate of density—and a far from trivial one—is the rate of damaging fights. Yet, rather than being the product of density per se, fights may have more to do with reduced escape opportunities. If so, instead of concluding that crowding makes monkeys aggressive, it would be more appropriate to say that once aggression does erupt, the consequences are more serious when space is limited.

In further support of the idea that primates actively cope with environmentally induced stress, reconciliation is more common under crowded conditions. There is a serious catch here, however: should we not correct for the number of fights? And should we not also correct for the general level of affiliation in the group? There simply is more opportunity to run into a former opponent accidentally in a cramped space. After we had applied these corrections, three specifically designed studies failed to confirm an increase in conciliatory tendencies under crowded conditions. Angeline van Roosmalen and I compared the reconciliation rate of the Arnhem chimpanzees indoors and outdoors, Josep Call did the same for rhesus monkeys in a corral and in a more crowded enclosure, and Filippo Aureli journeyed to Indonesia to see if long-tailed macaques in the rain forest reconciled as often as those in captivity. All three studies failed to demonstrate environmental effects on peacemaking tendencies.

One way to look at this finding is from the perspective of the social cage and the ties that bind individuals together. If the sense of belonging to a group, and the value of particular relationships, are similar across conditions, there is no reason to resolve fights more often in a corral or enclosure than in the wild or on an island. The alternative to peacemaking, namely avoidance between aggressor and victim, may not be a realistic option under any condition. Judge speculated that low-ranking rhesus monkeys on the island cannot leave their group for too long as they are probably better off being the last to feed among their group mates than competing on their own with strangers (the island is occupied by many different groups). Similarly, Aureli speculated that the risk of encountering large cats or birds of prey in the forest prevents monkeys from leaving the protection of the group even if they have serious problems with some of their group mates.[35]

The picture emerging from this research is that of a close-knit social system that produces x conflicts and requires y reconciliations for its maintenance regardless of whether it is held together by a moat, a fence, dangerous predators, or hostile groups. The tighter the perimeter, the more work needs to be put into keeping tensions down, partly because it is harder to escape physical damage if they do erupt.

It is well to realize, though, that coping with stress is by no means

WAR AND PEACE

Even in the closest group, conflicts are inevitable. One way to regulate and resolve them is through a clear-cut hierarchy. Nevertheless aggression may erupt, causing damage to valuable partnerships. To restore the peace, most primates engage in friendly reunions following fights.

Many animals have special signals to communicate who is dominant and who is subordinate. Two male chimpanzees demonstrate rank by creating an illusory size difference; the dominant walks upright with raised hair, the subordinate bows and pant-grunts. In reality, they are about the same size. *(Arnhem Zoo)*

As chimpanzees grow up, particularly the males, their youthful teasing gradually turns into serious challenges of the status quo. Here an adolescent swag-
gers in front of an adult female. In order to impress her, though, he still has to
grow a little. She lashes out, hitting him in the stomach. *(Yerkes Field Station)*

An adolescent male chimpanzee (right) teases his adult sister to the point that she cannot stand it any longer and bursts out screaming and gesticulating. *(Yerkes Field Station)*

Luit, Nikkie, and Yeroen (left to right) during a strategic reconciliation. The alpha male, Nikkie, has been fighting with his ally, Yeroen, providing their rival, Luit, with an opening to show off his strength. Without Yeroen's help, Nikkie is unable to restore order. Here he reaches out to Yeroen to mend their alliance before putting Luit in his place. *(From* Chimpanzee Politics; *Arnhem Zoo)*

Chimpanzees may not kiss exactly as we do, but the principle is the same. A kiss on the mouth (here by a female to a male) is this ape's conciliatory gesture par excellence. *(Yerkes Field Station)*

Browse food at the center of this scene has created conflict between the two female chimpanzees on the extreme left and right. As soon as one of them screamed, the alpha male, Jimoh, arrived with both arms spread as if beseeching them to stop quarreling. Despite his small size, Jimoh is an effective arbitrator of disputes. *(Yerkes Field Station)*

Even though bonobos and chimpanzees are closely related, they make peace in a strikingly different way. Bonobos do so via sexual and erotic contact, with partners of both the opposite sex and their own. One adult male offers a brief genital massage to another after a fight between them. Two adult females bare their teeth and squeal during genito-genital rubbing, and two juveniles engage in tongue-kissing (next page). *(San Diego Zoo)*

Savanna baboons during a greeting ritual in which one male tries to mount the other. *(Gilgil, Kenya)*

Two Chinese golden monkey females holding hands after a fight. One female opens her mouth widely, a conciliatory expression of this species. *(Courtesy of RenMei Ren; Peking University, Beijing)*

Capuchin monkeys make it clear when expectations have been violated. A juvenile male (right), trying to get a close look at an infant, is pushed away by the mother. The juvenile screams in protest; his tantrum is sustained for more than a minute. Immediately afterward, the juvenile spreads his legs and shows his erect penis. The mother responds to this typical appeasement posture by inspecting his genitals. *(Yerkes Primate Center)*

Kinship bonds provide a common context of conflict. A rhesus mother grabs the head of her pubertal daughter, who came too close to her new offspring (barely visible on the mother's belly). Her daughter grins submissively in response. *(Wisconsin Primate Center)*

One juvenile long-tailed macaque confronts another with an unmistakable threat face: open mouth, staring eyes, raised eyebrows, and spread ears. *(Photograph by Han van Beek; Courtesy of University of Utrecht)*

The hold-bottom gesture is characteristic of reunions and reconciliations among stump-tailed macaques. *(Yerkes Field Station)*

A free-ranging rhesus family on Morgan Island in South Carolina. The differences between captive and free-ranging monkeys are sometimes exaggerated. It is easier for monkeys to avoid attack where there is more space, but the sources of conflict never disappear. Social behavior and group organization have been found to be essentially the same under both conditions.

The elegant and elusive muriqui, or woolly spider monkey, may be one of the most peaceful primates on earth. *(Fazenda Montes Claros, eastern Brazil)*

the same as getting rid of it; constant behavioral (and probably also physiological) countermeasures are necessary under crowded conditions. All of these techniques are part of the impressive adaptive potential of the primate order. I often feel that the traditional approach of measuring how often primates fight or groom can only reveal the tip of the iceberg. If these measures show little variation, it may be because they do not adequately probe the modus vivendi each population arrives at after having lived for generations in a particular environment. The notion of negotiation and integration, of animals adjusting their intentions and expectations to those of others, is probably essential to an understanding of how social arrangements specific to the circumstances are agreed upon.

The outcome may parallel cultural variation. Human populations with long crowding histories, such as the Japanese, the Javanese, and the Dutch, each in their own way emphasize tolerance, conformity, and consensus, whereas populations spread out over lands with empty horizons may be more individualistic, stressing privacy and freedom instead. Inasmuch as the balance between individual and community values affects moral decisions, morality is an integral part of the human response to the environment, and an important counterweight to the social decay predicted for crowded conditions. Adjusting the definition of right and wrong is one of the most powerful tools at the disposal of *Homo sapiens,* a species of born adaptation artists. Morality is not the same during war and peace, or during times of plenty and scarcity, or under high or low population densities. If certain conditions persist for a long time, the entire moral outlook of a culture will be affected.

Even if the capacity to cope with social stress reaches its absolute peak in our species, it evidently has a long evolutionary history. Where it ultimately comes from can perhaps be answered by field-work on other primates. It is in the natural habitat that their behavior evolved, and no artificial environment, however well conceived, can replace the plants on which the species foraged for millions of years, the predators it tried to escape, the climate it had to deal with, and other environmental factors. If captive studies evaluate the adaptive potential of a species, fieldwork illuminates the adaptive significance of its behavior. *Adaptive significance* is defined as the advantages associated with a behavior that may have promoted the transmission of that behavior from one generation to the next until it became widespread in the species. Ideally, its study requires pristine ecosystems that have retained all their original qualities. These are hard to find nowadays because of worldwide pollution and habitat destruction, although it is generally assumed that current primate habitats are close enough to the hypothetical originals to at least shed *some* light on evolutionary adaptation.[36]

One such site is a remnant forest of 800 hectares in eastern Brazil, atop a mountain covered with coffee plantations. For the past ten years Karen Strier has raced up and down the ridges and valleys, cutting her way through an entangled understory, in order to keep up with muriquis effortlessly swinging through the canopy. These monkeys, weighing up to 15 kilograms, are the New World equivalent of the mountain gorilla: they are the largest indigenous primates on their continent—and one of the most endangered, with perhaps fewer than five hundred left in the wild. They are also true gentle giants.

One might object that we have heard false claims of peacefulness before, but in twelve hundred hours of observation, Strier counted a total of only nine aggressive chases among muriquis. Although she, like any primatologist, expected males near a sexually attractive female to compete with one another, she found that male muriquis simply wait their turn without any scuffles among them. Females have full control, writes Strier in *Faces in the Forest*: "I have often seen a female avoid a [genital] inspection by one male only to present herself with suggestive grins and twitters to another."[37] Similar sexual patterns and tolerance among males have been observed at another Brazilian site.

Males spend much time together, often embracing in clusters of two or more individuals while uttering a sound called the chuckle. The virtual absence of aggression is explained by Strier as a combination of separate foraging, egalitarian relations between the sexes, and the huge testicles of male muriquis. The last factor suggests that the arena of sexual competition may have shifted from direct contest over who gets to mate to the production of enough sperm to win the fertilization race against the sperm of other males. After copulation the female's vagina is blocked by a plug of fresh ejaculate, which is unceremoniously removed by the next male, to be dropped to the ground or eaten before he himself mates with her. Why things evolved in this direction is ambiguous, but it is this sort of field research that may reveal under which conditions evolution discards aggression as a conflict-settling strategy.

It is evident that the problem with which we started—that of aggression and violence in human society—cannot be solved by giving men bigger testicles. We have to work within our biological endowment, which is quite different from that of muriquis and rodents in that it includes both a rather combative temperament and powerful checks and balances.[38] Debates such as the one surrounding the Goodwin affair—about whether it is biology or environment that holds the key to society's woes—start from the erroneous assumption that the dividing line between the two is as clear-cut as that between the natural and social sciences.

If we accept that the biology of a species includes its typical behavior and adaptability, human biology has everything to do with environment: not only has it been shaped by past environments, it also determines how we respond to current environments. Insofar as we share these responses with other primates, study of their group life under a variety of conditions may help us see where the remarkable human plasticity comes from, and what it takes to stay out of the behavioral sink.

Community Concern

Socko, an adolescent chimpanzee at the Yerkes Field Station, teases Atlanta at his own peril. Atlanta may be fat and slow, but she makes up for it with an excellent memory. One moment, Socko throws dirt at her; more than an hour later, while Socko is absorbed in one of his exuberant games, we see Atlanta sneak up on him to grab an arm quickly and deliver a bite. Socko bursts out screaming, touching and squeezing the hurt body part. He runs to his "aunt" Peony and lets her take a look at it, all the while yelling indignantly. Peony gently takes the young male's arm and grooms him until he quiets down.

Still afraid of Atlanta, Socko walks in a wide arc around her to Rhett, her son. He tickles and wrestles with the juvenile, until both are rolling around over the ground uttering hoarse laughs. If it were not for the numerous quick glances he throws at Atlanta, one would think that Socko has forgotten the whole incident and is really enjoying himself. Instead, he is demonstrating good intentions and getting close to his opponent through contact with someone connected to her. Indeed, after a few minutes he moves from son to mother and sits down just out of reach of Atlanta, staring at something in the distance. Atlanta nods her head and shifts her position slightly in the way of a chimpanzee getting ready for a grooming session. Socko immediately responds by coming closer and picking through her hair. Soon Atlanta herself is grooming her tormentor.

Involvement of extra parties in the reconciliation process (Rhett in this instance) was first investigated by Peter Judge at the same field station but in a different species, the pigtail macaque. Like all macaques, pigtails form a matrilineal society in which female kin associate with and support one another. Apart from direct reconciliations between former combatants, Judge found that relatives of the victim tend to seek contact with the aggressor. A mother may approach the attacker of her daughter for what appears to be a reconciliation "on her offspring's behalf." If such so-called *triadic reconciliation* protects the victim's matriline against further hostilities, all of its members benefit, including the one who went to see the aggressor.[39]

Such may indeed be the case: tensions between individuals tend to generalize to their respective families. Cheney and Seyfarth noted in wild vervet monkeys that if the members of one matriline confront those of another, there is a reasonable chance that later in the day *other* members of the same matrilines will carry on the fight. These individuals must have watched the original incident and have developed hostile feelings toward the relatives of the monkeys who confronted their own kin. Triadic reconciliation, shown by vervets as well, is simply the other side of the same coin. Instead of getting even with a matriline that opposed their kin, they now try to mend relationships.

Reconciliation by proxy requires that monkeys know not just their own kin, but also to which kin group everyone else belongs. There are indeed growing indications that primates recognize the relationships around them. Crucial evidence has been produced by Verena Dasser, a Swiss primatologist. Showing macaques photographed portraits of adult females and juveniles of their group, she gave them the task of classifying the pictures on the basis of whether or not the depicted individuals were related to each other. Her subjects handled the task well, demonstrating that, like Socko, they recognized the mother-child connection.[40]

Knowledge of relationships in which they are not directly involved explains why fights among primates do not always remain restricted to the parties who started them, and why peacemaking may ripple well beyond a conflict's epicenter. Some species, such as the stumptailed macaque, even utter special squeals during reconciliation that alert and attract the entire group to what is going on. Yet there is probably no nonhuman primate that keeps a closer eye on reconciliations, and has a better grasp of their significance, than the chimpanzee. I have even seen members of this species react *negatively* when one among them offered an olive branch to another.

It is not unusual for females in a captive chimpanzee colony to band together to protect themselves against an abusive male. Given that such female coalitions can deliver quite a beating, the male is in an understandable hurry to get out of their way. He watches the other sex from a safe distance if he has been lucky enough to escape. Because none of the females matches him in strength and speed, their solidarity is crucial. During these standoffs, I have seen females turn all their pent-up fury against the first among them to approach the adversary to present her behind, or kiss him. Apparently, the other females were not finished with him yet and saw this one female's peace initiative as treason.

Much more commonly, though, responses to reconciliation are positive, even celebratory. After a major confrontation, juvenile chimpanzees may hang around unreconciled adult males, looking from the one to the other, only to jump with loud hooting on the males' backs as soon as they hug each other. And, as described before, adult females sometimes serve a catalytic function by bringing rival males together, an intervention that requires quite a degree of social awareness. Such mediation demonstrates the value chimpanzees attach to harmonious coexistence, as also reflected in the pandemonium and embracing by an entire colony following a dramatic reconciliation in its midst.

Triadic reconciliation, third-party mediation, and mass excitement about evaporated tensions suggest that monkeys and apes care about the state of relationships in their group or community. And not just their own private relationships, but also those around them. I have labeled this *community concern*. It is not that I believe monkeys and apes worry about the community as an abstract entity; they seem rather to strive for the kind of community that is in their own best interest. Insofar as the interests of different individuals overlap, community concern is a collective matter.

If a female chimpanzee, for example, confiscates the heavy rock in the hand of a male who is about to challenge a rival, or acts as go-between after an actual confrontation between the two, the whole group benefits even if she did so only for her own sake. It is not hard to see what she stands to gain by improved male relations. A group of savagely brutal and continuously edgy males poses a grave threat to every female and the offspring she is trying to raise. Moreover, in the wild, such males would have trouble presenting a united front to the outside, thus inviting raids and territorial takeovers by neighboring males.

If individual interests were to overlap completely, everything would be straightforward; there would be total agreement on the value of social harmony. But individuals have clashing interests as well. A young male's climbing of the rank order may stir up a lot of trouble, throwing the group into chaos for weeks or months on end. Or a matriarch may seek confrontation after confrontation with members of a higher-ranking matriline weakened by the death of its own matriarch. There is high tension between private goals relating to status and access to resources and common goals relating to the group's success in its particular ecological setting. When individual members jockey for the best social position, or for the largest share of food, the wider network of relations and the group as a whole may suffer.

This situation is dangerous, given the reliance of social primates on each other for defense against outside threats and for finding food and water. They can ill afford to be at war with the companions on whom they depend. Community concern takes as its starting point each organism's vested interest in a social environment optimal for survival and reproduction. The drawbacks of competition can be summarized as follows.[41]

Harm to specific partnerships: Social primates face the dilemma that they sometimes cannot win a fight without losing a friend. Profitable partnerships need to be nursed and protected, a process that requires the suppression of aggressive outbursts. In a few species these inhibitions are so powerful that resources are peacefully shared instead of generating competition. Occasionally even the best relationships explode, as described for the ruling alliance in Arnhem between Nikkie and Yeroen. Reconciliation serves to contain the damage.

Bodily harm to others: Companions who are limping, bleeding, and suffering from infections are of little use against predators or neighboring groups. Hyperaggressive individuals who cause frequent injuries compromise the quality of the social environment on which their very survival depends.

Harm to group unity: Given the value of the group, preservation of social cohesion serves each and every member. Exceptions do occur when individuals immigrate into a group and disrupt it by efforts to attain an advantageous position, or when resident members start a protracted dominance struggle. The payoffs of such activities are evidently high enough to outweigh social disruption. In the long run, however, members who regularly and seriously disrupt group life damage—along with the interests of everyone else—the interests of themselves and their relatives in the group.

Community concern can be defined, then, as *the stake each individual has in promoting those characteristics of the community or group that increase the benefits derived from living in it by that individual and its kin*. This definition does not hinge on conscious motives and intentions; it merely postulates benefits associated with particular behavior. It says that social animals have been selected to inhibit actions that may destroy group harmony and to strive for an optimal balance between peaceful coexistence and the pursuit of private interests. Whether animals realize how their behavior impacts the group as a whole is not critical for the evolution of community concern any more than that it is critical that animals know how sex relates to reproduction in order to pursue optimal reproductive strategies.

This said, motives and intentions are fascinating to contemplate, particularly because human morality can be looked at as community concern made explicit to the fullest degree. The higher a species' level of social awareness, the more completely its members realize how events around them ricochet through the community until they land at their own doorstep. This understanding allows them to become actively involved in shaping community characteristics. It starts with interest in relationships close to them (as between their kin and an attacker), then extends to more distant relationships that impact their lives (strained relations between rival dominants), and culminates in collective support for actions that enhance group harmony (arbitration by a central individual, who performs the so-called control role).

Not satisfied with a society fashioned by uncoordinated individual efforts, one of humanity's chief accomplishments is to translate egocentric community concerns into collective values. The desire for a modus vivendi fair to everyone may be regarded as an evolutionary outgrowth of the need to get along and cooperate, adding an ever-greater insight into the actions that contribute to or interfere with this objective. Our ancestors began to understand how to preserve peace and order—hence how to keep their group united against external threats—without sacrificing legitimate individual interests. They came to judge behavior that systematically undermined the social fabric as "wrong," and behavior that made a community worthwhile to live in as "right." Increasingly, they began to keep an eye on each other to make sure that their society functioned in the way they wanted it to function.

Conscious community concern is at the heart of human morality.

6

CONCLUSION

I question whether the spiritual life
does not get its surest and most ample
guarantees when it is learned that the
laws and conditions of righteousness
are implicated in the working processes of the universe; when it is found
that man in his conscious struggles, in
his doubts, temptations and defeats, in
his aspirations and successes, is moved
on and buoyed up by the forces which
have developed nature.

John Dewey[1]

Even if animals other than ourselves act in ways tantamount to moral behavior, their behavior does not necessarily rest on deliberations of the kind we engage in. It is hard to believe that animals weigh their own interests against the rights of others, that they develop a vision of the greater good of society, or that they feel lifelong guilt about something they should not have done.

What Does It Take to Be Moral?

Members of some species may reach tacit consensus about what kind of behavior to tolerate or inhibit in their midst, but without language the principles behind such decisions cannot be conceptualized, let alone debated. To communicate intentions and feelings is one thing; to clarify what is right, and why, and what is wrong, and why, is quite something else. Animals are no moral philosophers.

But then, how many *people* are? We have a tendency to compare animal behavior with the most dizzying accomplishments of our race, and to be smugly satisfied when a thousand monkeys with a thousand typewriters do not come close to William Shakespeare. Is this a reason to classify ourselves as smart, and animals as stupid? Are we not much of the time considerably less rational than advertised? People seem far better at explaining their behavior after the fact than at considering the consequences beforehand. There is no denying that we are creatures of intellect; it is also evident that we are born with powerful inclinations and emotions that bias our thinking and behavior.

A chimpanzee stroking and patting a victim of attack or sharing her food with a hungry companion shows attitudes that are hard to distinguish from those of a person picking up a crying child, or doing volunteer work in a soup kitchen. To classify the chimpanzee's behavior as based on instinct and the person's behavior as proof of moral decency is misleading, and probably incorrect. First of all, it is uneconomic in that it assumes different processes for similar behavior in two closely related species. Second, it ignores the growing body of evidence for mental complexity in the chimpanzee, including the possibility of empathy. I hesitate to call the members of any species other than our own "moral beings," yet I also believe that many of the sentiments and cognitive abilities underlying human morality antedate the appearance of our species on this planet.

The question of whether animals have morality is a bit like the question of whether they have culture, politics, or language. If we take the full-blown human phenomenon as a yardstick, they most definitely do not. On the other hand, if we break the relevant human abilities into their component parts, some are recognizable in other animals (see page 211).

Culture: Field primatologists have noticed differences in tool use and communication among populations of the same species. Thus, in one chimpanzee community all adults may crack nuts with stones, whereas another community totally lacks this technology. Group-specific signals and habits have been documented in bonobos as well as chimpanzees. Increasingly, primatologists explain these differences as learned traditions handed down from one generation to the next.[2]

Language: For decades apes have been taught vocabularies of hand signals (such as American Sign Language) and computerized symbols. Koko, Kanzi, Washoe, and several other anthropoids have learned to effectively communicate their needs and desires through this medium.

It is hard to imagine human morality without the following tendencies and capacities found also in other species.

Sympathy-Related Traits
Attachment, succorance, and emotional contagion.
Learned adjustment to and special treatment of the disabled and injured.
Ability to trade places mentally with others: cognitive empathy.*

Norm-Related Characteristics
Prescriptive social rules.
Internalization of rules and anticipation of punishment.*

Reciprocity
A concept of giving, trading, and revenge.
Moralistic aggression against violators of reciprocity rules.

Getting Along
Peacemaking and avoidance of conflict.
Community concern and maintenance of good relationships.*
Accommodation of conflicting interests through negotiation.

* It is particularly in these areas—empathy, internalization of rules and sense of justice, and community concern—that humans seem to have gone considerably further than most other animals.

Politics: Tendencies basic to human political systems have been observed in other primates, such as alliances that challenge the status quo, and tit-for-tat deals between a leader and his supporters. As a result, status struggles are as much popularity contests as physical battles.

In each of these domains, nonhuman primates show impressive intelligence yet do not integrate information quite the way we do. The utterances of language-trained apes, for example, show little if any evidence of grammar. The transmission of knowledge from one generation to the next is rarely, if ever, achieved through active teaching. And it is still ambiguous how much planning and foresight, if any, go into the social careers of monkeys and apes.

Despite these limitations, I see no reason to avoid labels such as "primate culture," "ape language," or "chimpanzee politics" as long as it is understood that this terminology points out fundamental similarities without in any way claiming *identity* between ape and human behavior. Such terms serve to stimulate debate about how much or little animals share with us. To focus attention on those aspects in which we differ—a favorite tactic of the detractors of the evolutionary perspective—overlooks the critical importance of what we have in common. Inasmuch as shared characteristics most likely derive from the common ancestor, they probably laid the groundwork for much that followed, including whatever we claim as uniquely ours. To disparage this common ground is a bit like arriving at the top of a tower only to declare that the rest of the building is irrelevant, that the precious concept of "tower" ought to be reserved for the summit.

While making for good academic fights, semantics are mostly a waste of time. Are animals moral? Let us simply conclude that they occupy a number of floors of the tower of morality. Rejection of even this modest proposal can only result in an impoverished view of the structure as a whole.

Floating Pyramids

It is hard to take care of others without taking care of oneself first. Not that people need a mansion and a fat bank account before they can be altruistic, but certainly we do not expect much assistance from someone in poor health without the most basic means of subsistence. Paradoxically, therefore, altruism starts with an obligation to oneself.

The form of altruism closest to egoism is care of the immediate family. In species after species, we see signs of kin selection: altruism is disproportionally directed at relatives. Humans are no exception. A father returning home with a loaf of bread will ignore the plight of whomever he meets on his path; his first obligation is to feed his family. This pattern of course says nothing about the inherent value of his own children compared to the others living in the neighborhood. If his family were well fed and everybody else were starving, it would be a different matter—but if his family is as hungry as the rest, the man has no choice.

The circle of altruism and moral obligation widens to extended family, clan, and group, up to and including tribe and nation. Benevolence decreases with increasing distance between people. Going against the grain of this natural gradient meets with sharp disapproval. Spies are despised precisely because they help an out-group at the expense of the in-group. Similarly, we are shocked that people under the East German communist regime informed on parents and spouses, putting nation before family. And if the father in the above example had come home empty-handed because of sympathy for strangers, his family would have shown very little understanding.

Altruism is bound by what one can afford. The circle of morality reaches out farther and farther only if the health and survival of the innermost circles are secure. For this reason, rather than an expanding circle I prefer the image of a floating pyramid. The force lifting the pyramid out of the water—its buoyancy—is provided by the available resources. Its size above the surface reflects the extent of moral inclusion. The higher the pyramid rises, the wider the network of aid and obligation. People on the brink of starvation can afford only a tiny tip of the moral pyramid: it will be every man for himself. It is only under the most extreme conditions, however, perhaps like those described for the Ik by Colin Turnbull, that such "lifeboat ethics" apply.

As soon as the immediate threat to survival is removed, members of our species take care of kin and build exchange networks with fellow human beings both inside and outside their group. Compared

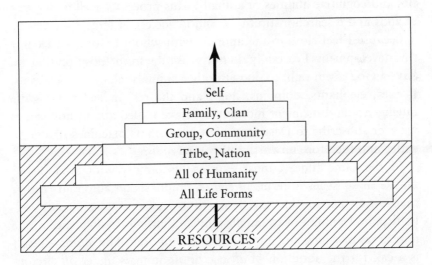

The expanding circle of human morality is actually a floating pyramid. Altruism is spread thinner the farther away we get from our immediate family or clan. Its reach depends on resources and affordability; the pyramid's buoyancy determines how much of it will emerge from the water. The moral inclusion of outer circles is thus constrained by obligations to the inner ones.

to other primates, we are a remarkably giving species. Moral inclusion does not imply, though, that every person is valued exactly the same. In principle they may be equal, but in practice human kindness and cooperativeness are spread thinner the farther we get from kin and community.

The ideal of universal brotherhood is unrealistic in that it fails to distinguish between these innermost and outermost circles of obligation. The American human ecologist Garrett Hardin disdainfully refers to indiscriminate kindness as "promiscuous altruism." If altruism evolved because of a need to cooperate against hostile forces, solidarity with what is close against what is distant is an integral part. As observed by the French anarchist Pierre-Joseph Proudhon more than a century ago: "If everyone is my brother, I have no brothers."[3]

Depending on what a society can afford, then, the moral pyramid may swell to giant size, in principle embracing all of humanity, but always retaining its fundamental shape. Life forms other than our own may be included. Recent studies of animal behavior, mine included, provide ample reason to reconsider the way animals are used for science, entertainment, food, education, and other purposes. We need to re-evaluate traditional attitudes developed over a long history without realistic alternatives, and without awareness of the sensibilities and cognitive abilities of animals. This process is well under way at zoos and research institutions, and in society at large.

Because I feel close to the animals with whom I work, I welcome this development. I certainly do not subscribe to the position that we have a God-given right to do with other animals whatever pleases us. If apes, elephants, dolphins, dogs, and the rest indeed possess the intelligence and incipient morality we have talked about, how could we ever subscribe to Descartes' view of them as machines unable to suffer and therefore unworthy of compassion?

At the same time, I must express discomfort with attempts to phrase these issues in terms of *rights*. Emphasis on autonomy rather than on connection has given rise to a discourse that is cold, dogmatic, and leaning toward an absolutism that fails to do justice to the gray areas of which human morality is composed. The ultimate result is a call for the abolition of *all* use of *all* animals under *all* circumstances, from hunting to meat consumption, from keeping them in zoos to having them work on the farm. In the process, we sometimes ignore our first obligation, which is to fellow human beings.

A particularly radical proposal is that of Paola Cavalieri and Peter Singer. Together with a number of prominent scientists, in a volume entitled *The Great Ape Project,* they advocate a "community of equals" consisting of apes and humans. They see no good reason why animals as close and similar to us as the great apes should fall into a different moral category. Why not elevate them to the same legal status as their bipedal relatives?

The logical flaw in this proposal is its blatant anthropocentrism. How can one make similarity to a particular species the touchstone of moral inclusion without ranking that species above the rest? If rights increase in proportion to the number of humanlike characteristics possessed by a species, it is hard to escape the conclusion that humans themselves deserve the most rights of all.

A second problem is that rights are normally accompanied by responsibilities, which cannot possibly apply to apes. The authors reply that since mentally retarded people are exempt from this linkage, why not apes?

To my mind, Cavalieri and Singer's plea reflects profound condescension. Have we really reached the point at which respect for apes is most effectively advocated by depicting them as retarded people in furry suits? And while we are at it, why should we not then classify a baboon as a mentally challenged ape? It seems endless: once apes are granted equal status on such questionable grounds, there is no way to keep out cockroaches. My own feeling is that we must take the inherent beauty and dignity of animals as our starting point.

No matter how well intentioned the concerns of animal rights advocates, they are often presented in a manner infuriating to anyone concerned about *both* people and animals. Human morality as we know it would unravel very rapidly indeed if it failed to place human life at its core. Again, there is no judgment here about the objective value of our lives compared to the lives of other creatures. Personally, I do not feel superior to a butterfly, let alone to a cat or a whale. But who can deny our species the right to construct its moral universe from a human perspective?

It will be up to society to decide whether it will continue to support certain kinds of research on certain kinds of animals. It is already common practice in biomedical research that if a particular experiment on monkeys is considered no more effective than on rats, the monkey study will never be conducted. Similarly, if an experiment on chimpanzees is judged no more effective than one on monkeys, the first study will simply not take place.

Unfortunately for the animals, they are not the only ones hanging in the balance. Human lives are also at stake. Anyone who enters a hospital or picks up a prescription at the pharmacy makes use of animal testing. Few people consider it trivial to fight diseases such as AIDS, that affect millions. If a vaccine could be developed without using animals, of course that would be preferable. But there are no signs that this stage will be reached anytime soon. Choices must be made, and these get more difficult the more complex the life forms serving as guinea pigs.

How much do we care and what can we afford? There are excellent reasons to insist on respect and concern for animals that serve the human cause. Apes do warrant special consideration. We should either phase out experiments on certain species altogether, or if humanity cannot forgo the benefits derived from them, we must at least enrich and enhance their lives in captivity and reduce their suffering. Phrasing the issue, as I do here, in terms of our *responsibility* to other life forms leaves the moral pyramid intact, and may lead to less radical conclusions than phrasing it in terms of rights. All the same, it is no easier to resolve the dilemmas facing us.[4]

A Hole in the Head

On September 13 of 1848 Phineas Gage, while leveling terrain for a railroad track in New England, suffered a hideous accident that would make him a neurological *cause célèbre*. Owing to a momentary distraction, Gage triggered a blast while leaning over a hole filled with explosive powder. The pointed tamping iron that he held in his hands was hurled like a rocket straight through his left eye, brain, and skull. Incredibly, Gage was only briefly stunned. He instantly regained consciousness and was able to walk and talk immediately afterward. The meter-long iron lay in the sand, meters away.

The twenty-five-year-old foreman recovered completely, retained all elementary mental functions, and remained able-bodied for the rest of his life. His speech was normal, he absorbed new information as before, and he showed no lapses of memory. However, his personality changed. From a pleasant and reliable fellow, popular among his peers, he turned into someone who could not hold a job because he had lost all respect for social conventions. He would lie and curse uncontrollably. Perhaps the greatest change was that his sense of responsibility vanished: he could not be trusted to honor commitments. According to his physician, the equilibrium between intellectual faculties and lower impulses had been disturbed by the accident.

THE FAR SIDE   By GARY LARSON

© Chronicle Features, 1981 *Larson* 12-2

"Fair is fair, Larry . . . We're out of food, we drew
straws — you lost."

Lifeboat ethics with dog (1981 The Far Side cartoon by Gary Larson is reprinted by permission of Chronicle Features, San Francisco, California. All
rights reserved.)

The neurologist Hanna Damasio and her coworkers recently reported on an inspection of Gage's skull and the tamping iron—both
preserved in a museum at Harvard University. They made computer
models of the brain damage. Apparently the transformation from an
upright citizen into a man with serious character flaws had been
brought about by lesions in the ventromedial frontal region of his
brain. This pattern fits that of a dozen other brain-damaged patients
known to science who have intact logical and memory functions but
compromised abilities to manage personal and social affairs. It is as
if the moral compass of these people has been demagnetized, causing
it to spin out of control.

What this incident teaches us is that conscience is not some disem-
bodied concept that can be understood only on the basis of culture
and religion. Morality is as firmly grounded in neurobiology as anything else we do or are. Once thought of as purely spiritual matters,

honesty, guilt, and the weighing of ethical dilemmas are traceable to specific areas of the brain. It should not surprise us, therefore, to find animal parallels. The human brain is a product of evolution. Despite its larger volume and greater complexity, it is fundamentally similar to the central nervous system of other mammals.

We seem to be reaching a point at which science can wrest morality from the hands of philosophers. That this is already happening—albeit largely at a theoretical level—is evident from recent books by, among others, Richard Alexander, Robert Frank, James Q. Wilson, and Robert Wright. The occasional disagreements within this budding field are far outweighed by the shared belief that evolution needs to be part of any satisfactory explanation of morality.

Gardener and garden are one and the same. The fact that the human moral sense goes so far back in evolutionary history that other species show signs of it plants morality firmly near the center of our much-maligned nature. It is neither a recent innovation nor a thin layer that covers a beastly and selfish makeup.

It takes up space in our heads, it reaches out to fellow human beings, and it is as much a part of what we are as the tendencies that it holds in check.

NOTES

Prologue

1. Huxley, 1989 (1894), p. 83.
2. Williams, 1988, p. 438.
3. Dewey, 1993 (1898), p. 98.

1. Darwinian Dilemmas

1. Dawkins, 1976, p. 3.
2. Gould, 1980, p. 261.
3. Dettwyler, 1991, p. 382.
4. Kurland, 1977, p. 81.
5. Midgley, 1991, p. 8.
6. Wilson, 1975, p. 562.
7. According to Kenneth Lux, opposition to welfare assistance (the so-called Poor Laws) was most evident in the second edition of Malthus' *Essay on the Principle of Population* and was expunged from subsequent editions: "A man who is born into a world . . . if he cannot get subsistence from his parents on whom he has a just demand and if the society does not want his labour, has no claim of right to the smallest portion of food, and, in fact, has no business to be where he is. At nature's mighty feast there is no vacant cover for him. She tells him to be gone,

and will quickly execute her own orders, if he does not work upon the compassion of some of her guests. If these guests get up and make room for him, other intruders immediately appear demanding the same favour" (quoted in Lux, 1990, pp. 34–35).

8. Rockefeller quoted in Lux, 1990, p. 148.

9. History is not as simple as presented here. Charles Darwin, Alfred Russell Wallace, Thomas Henry Huxley, and Herbert Spencer each took a different position with regard to the (im)possibility of an evolved morality. Well-documented accounts of this early debate may be found in Richards (1987) and Cronin (1991). See also Nitecki and Nitecki (1993).

10. Yerkes and Yerkes, 1935, p. 1024.

11. Gene-centric sociobiologists often speak of "a gene for behavior x," regardless of what is known about the heritability of behavior x (usually, little or nothing). In reality, each gene acts in conjunction with hundreds of others. So every behavior is likely to depend on a wide range of genetic factors. Even if we grant gene-centric sociobiologists that their one-gene–one-behavior scheme is not to be taken literally—that it is a mere shorthand for discussion—it is advisable to balance it with another generalization, one that is at least as close to the truth: "Every character of an organism is affected by all genes and every gene affects all characters" (Mayr, 1963, p. 164).

12. Apparently Dawkins is not convinced that we are born selfish, in the vernacular sense. In response to Midgley (1979) he admits that selfish-gene rhetoric may well be out of touch with actual human motives: "To the extent that I know about human psychology (a rather small extent), I doubt if our emotional nature is, as a matter of fact, fundamentally selfish" (Dawkins, 1981, p. 558).

This is a message to bear in mind, for it certainly is not evident in the author's writings. A general problem with pop sociobiology is that complex issues are compressed to such a degree that even if the author is fully aware of what is left out, the reader has no way of knowing. The simplifications are then perpetuated ad nauseam by less-informed writers until they haunt the field in general and must be countered as if they represented serious ideas (Kitcher, 1985).

In *The Ethical Primate*, Midgley (1994, p. 17) has reiterated her views on the pitfalls and illusions of reductionist science, giving scathing attention to sociobiology's forays into the psychological domain: "Darwinism is often seen—and indeed is often presented—not as a wide-ranging set of useful suggestions about our mysterious history, but as a slick, reductive ideology, requiring us, in fact, to dismiss as illusions matters which our experience shows to be real and serious."

13. Hamilton, 1971, p. 83; Dawkins, 1976, p. 215.

14. Williams, 1989, p. 210.

15. Williams, 1989, p. 210. One may wonder if Williams really meant to condemn Mother Nature as a wicked old witch. Perhaps, instead of *immoral,* he intended to say that nature is *amoral,* which is of course exactly what Huxley meant by "morally indifferent." However, Williams makes it quite clear that he sees a contrast between the biological and the physical order: "I would concede that moral indifference might aptly characterize the physical universe. For the biological world a stronger term is needed" (p. 180). When he distinguishes being struck by lightning (physical process) from being struck by a rattlesnake (animal action), Williams calls the behavior of the snake and other animals "grossly immoral." In normal usage this judgment would imply disapproval, yet he does not believe that an animal can be held accountable for its actions. But since there can be no morality without individual responsibility, Williams selected the wrong term: he does mean that nature is amoral, and his whole tirade unravels.

16. NSF Task Force, Newsletter of the Animal Behavior Society, vol. 36 (4).

17. Frank, 1988, p. 21.

18. The only similar report that I am aware of concerns captive dolphins. Two females showed interest in the labor of a third, remaining close to her until the fetus was expelled. The older of the two attending females and the mother then swam under the baby dolphin, one on each side. Had the infant not been able to reach the surface by itself, the two females most likely would have lifted it between their dorsal fins (McBride and Hebb, 1948).

19. Gould, 1988, title.

20. Kropotkin, 1972 (1902), pp. 18, 59.

21. To say, as Lorenz (1966) did, that animals rarely kill members of their own species because then the species would die out, assumes that animals care about the well-being of their group or species. Such naive group selectionism was dismissed by Williams (1966), who argued that variants pursuing this goal would rapidly lose out to variants placing private interests first. Natural selection favors individuals who procreate more successfully than others; the interests of groups or species are relevant only insofar as they overlap with those of individuals.

   Extreme sacrifices, however, such as human warriors endangering or giving their lives in combat, pose a serious challenge to this line of thought. Do these warriors not place the good of their group above private interests? To explain their behavior, it has been speculated that status and privilege accrue to surviving heroes, or to the families of those who actually lost their lives. If this is indeed the case, heroic acts on behalf of the community may increase the warrior's reproduction or the survival of his offspring, an argument attributed to R. A. Fisher by Alexander (1987, p. 170). Note, however, how this explanation injects moral mechanisms, such as approbation and gratitude, into a discussion about the origin of morality, creating a rather circular argument. Furthermore, it is hard to believe that, in practice, the families of fallen soldiers are better off than the families of soldiers who return alive from battle.

22. Pronouncements of the demise of group selection theory have been premature. Selection at the level of groups probably operates along with selection at the level of individuals and genes. Such "nested" selection models by no means introduce noncompetitive principles; rather, they transpose conflict up one level, from individual against individual to group against group (Wilson, 1983; Wilson and Sober, 1994).

23. Darwin, 1981 (1871), vol. 1, p. 166.

24. De Mandeville, 1966 (1714), pp. 18–24.

25. Smith, 1982 (1776), bk. 3, p. 423.

26. Smith, 1937 (1759), p. 9.

27. Ethologists distinguish sharply between proximate and ultimate causes. *Proximate causes* concern learning, experience, and the direct circumstances and motivations underlying behavior. *Ultimate causes* promoted a behavior in the course of evolution. If a behavior assists survival and reproduction, for example because it repels predators or attracts mates, this is the ultimate reason for its existence. Since evolution takes place on a timescale that escapes perception, only proximate causes exist in the minds of animals and most humans. Students of evolutionary biology are unique in that they care about ultimate causes.

    Unfortunately, proximate and ultimate levels are frequently confused, particularly when the function of a behavior seems so obvious that it is hard to imagine that the actors are oblivious to it. Popular nature documentaries contribute to the mixup by describing animal behavior in ultimate terms. They will explain that two male walruses fight over the right to impregnate a female, whereas these males neither know nor care about what happens in the female's womb after they have mated.

28. De Waal and van Roosmalen, 1979, p. 62.

29. Once, after I had explained these theories to a political scientist with antisociobiological sentiments, he commented with some *Schadenfreude,* "Oh, but then you are getting into exactly the same mess we are in." He meant that instead of having the neat, crisp, reductionist picture of human behavior advertised by the early sociobiologists, we are introducing so many layers and refinements that the complexity may begin to overwhelm us as does the hodgepodge of theories confronting the social sciences. The big difference, of course, is that biologists have a single core theory within which everything must somehow make sense, whereas the social sciences lack such an integrative framework.

30. The first use of the word "ethology" in its current meaning was a reaction against the laboratory-based biological science of the influential Baron Cuvier. Cuvier's most important adversary in the debates at the

Académie des Sciences was Etienne Geoffroy-Saint-Hillaire, the father of Isidore, who proposed the ethology label. The term referred to the study of animals as living beings in nature, as opposed to the Cuvierian cadavers that smelled of formaldehyde. At approximately the same time, however, the renowned German evolutionist Ernst Haeckel coined *Ökologie* (which became the English "ecology") for the relation between the organism and its environment. This term immediately overshadowed "ethology" and generated confusion about the exact meaning of the latter. Jaynes (1988) believes that the closeness in meaning, combined with the association of early French ethology with Lamarckism, prevented ethology from developing into a significant movement in the nineteenth century.

31. Age-specific symbol-learning sensitivity may extend to nonhuman primates. When Sue Savage-Rumbaugh tried to teach symbols to a fully adult bonobo, she met with little success. The ape, despite being cooperative and bright, learned only seven symbols; her two-and-a-half-year-old son, on the other hand, learned from just sitting in on the training sessions. Without instruction or reward, he picked up the use of many symbols and comprehended hundreds of spoken English words (Savage-Rumbaugh et al., 1986).

32. Special learning abilities or sensitivities involved in the acquisition of moral consciousness have also been discussed by Lewin (1977), Simon (1990), and Wilson (1993, pp. 148–152).

33. Well-known ethologists, such as Wolfgang Wickler, Irenäus Eibl-Eibesfeldt, and Konrad Lorenz, have extensively speculated in popular books about the biological roots of human ethics. It must have been increased awareness of the naturalistic fallacy that compelled Wickler (1981) to add a subtitle to the second edition of his best-seller, *Die Biologie der Zehn Gebote* (The biology of the Ten Commandments), which literally reads *Warum die Natur für uns kein Vorbild ist* (Why nature does not serve us as example). This literature was critically reviewed by the late German anthropologist and primatologist Christian Vogel (1985, 1988).

34. The fatal incident in the Arnhem colony was interpreted as political murder. It resulted from a collapse of the ruling coalition because of the leader's failure to grant sexual privileges to his ally. In the resulting power vacuum, another male suddenly rose to the top. He paid for this ten weeks later when the two frustrated former allies banded together at night to injure him so badly that his life could not be saved (de Waal, 1986a; 1989a, pp. 59–69).

Since this was the very first report of such severe fighting within an established group, it may be tempting to dismiss it as a product of captivity. Recently, however, a similar event was described by Goodall (1992) for wild chimpanzees. The reigning alpha male fell from power after a gang attack that resulted in serious damage to his scrotum (the ensuing infection might have killed him had it not been for veterinary intervention). This intracommunity aggression was by far the most savage observed during thirty years at Gombe; such belligerence is more typical between communities (Goodall, 1986).

35. Lorenz, 1966 (1963), p. 167.

36. Hume, 1978 (1739), p. 469.

37. A wide range of views exists on biological constraints on morality. My personal opinion is that the evolutionary process provided us with the ability and the prerequisites for morality, as well as with a set of basic needs and desires that morality needs to take into account. The moral decisions themselves, however, are left to be negotiated among the members of society, hence are by no means specified by nature. Ruse (1986), in contrast, believes that "ought" feelings, such as a felt duty to assist others, have been put in place directly by natural selection: "We are talking of more than a mere feeling that we want to help others. It will be an *innately based sense of obligation* towards others" (p. 222; italics added). These differences of opinion need to be worked out within the framework of evolutionary ethics, the basic tenet of which is that the moral sense is not antithetical to but an integrated part of human nature (Ruse, 1988; Wilson, 1993).

2. Sympathy

1. Darwin,, 1981 (1871), vol. 1, pp. 71–72.

2. Wispé, 1991, p. 80.

3. On the basis of this incident, Porter (1977, p. 10) comments that he would not automatically discount the numerous reports of people who claim to have been saved by porpoises or some of the smaller whales. Accounts of Cetacea helping humans generally describe one of the following: (a) a drowning person is lifted to the surface; (b) a boat or ship is guided to safety (around submerged rocks, out of a storm); or (c) a swimmer is protected against sharks by a dolphin cordon. Reviews of both interspecific and intraspecific succorant behavior of Cetacea may be found in Caldwell and Caldwell (1966), Connor and Norris (1982), and Pilleri (1984).

4. Porter, 1977, pp. 10, 13.

5. Eibl-Eibesfeldt, 1990, p. 156.

6. Did Yeroen intentionally manipulate the other male's perception, or had he simply learned that limping reduces the risk of attack? The first possibility would have required him to imagine how he himself looked from the other's perspective; the second would have required little else than a rewarding experience during a period when he had been limping of necessity.

   Even if it is increasingly believed that great apes possess a capacity for intentional deception, it is impossible to prove this capability in each instance. Moreover, the false distress described for a female gorilla (Hediger, 1955, pp. 150–151) and for Yeroen (de Waal, 1982, pp. 47–48) is paralleled by similar occurrences in species supposedly lower on the cognitive scale. Dog owners tell me of faked limps by pets trying to get attention, and Caine and Reite (1983, p. 25) describe a female macaque with signs of malingering: "Whenever she was placed in her social group she limped badly, although, upon examination, no evidence of injury or disease was found. Furthermore, the limping disappeared when the animal was housed alone."

This female's behavior may have had more to do with how veterinarians treat limping monkeys than with the reaction of conspecifics. Perhaps preferring to be alone, she may have learned from a real physical trauma in the past how to get people to remove her from the group. If so, Yeroen's deception differed from this monkey's, as well as from that of the gorilla and the dogs mentioned previously, in at least one aspect: it was intended to appeal to members of his own species.

7. Lieberman, 1991, p. 169.
8. Play inhibitions obviously are not limited to primates. As the owner of any large dog can testify, they are even more dramatic in animals such as carnivores that are equipped to do horrible damage in a fraction of a second. The acquisition of such inhibitions was observed in two female black bear cubs, Kit and Kate, raised by Ellis Bacon for a wildlife project in the Smoky Mountains. Bear cubs, although very appealing, can be quite aggressive in play and enter wrestling matches with an energy and force that totally overwhelm a human partner. Human skin being paper-thin compared to that of a bear, the legs and arms of the cubs' caretakers were covered with scratches, bites, and bruises. A dramatic, welcome change in play style occurred, however, when the cubs were approximately eight months old: "They acted as if they discovered he [Bacon] was different from a bear: there was a rapid decline of inadvertent scratching and biting, and they distinguished clothing (which was still fair game) from flesh. From then on, with the exception of occasional dominance testing, Ellis considered himself physically safe in their presence. In contrast, the cubs were still very rough with each other in play or fights" (Burghardt, 1992, p. 375).

Moss (1988, p. 163) describes how the world's most formidable play partner, a bull elephant, learned to downsize in order to have fun: "Earlier in the year I had seen a large adult bull, Mark, lie down in an upright position in order to spar with another bull who was considerably smaller than he was. They had sparred playfully and briefly with both standing, but the young bull, M140, turned away, and although Mark followed, M140 would not spar with him again. Then Mark sank down on his knees with his rear legs out behind him, and as soon as M140

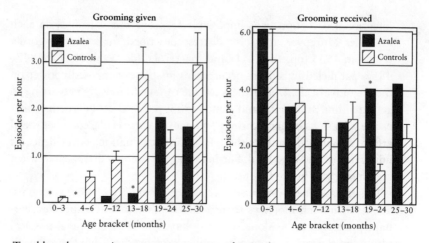

Total hourly grooming rates per age group for Azalea, a trisomic rhesus monkey, and the mean (plus standard error of mean, SEM) for twenty-three normal female peers of the same age (controls). Asterisks denote periods in which Azalea ranked at the extreme top or bottom of the distribution.

saw him he came straight over and started sparring. M140 was now the taller of the two."

9. After having devised this example, I heard from Sue Savage-Rumbaugh a striking story about an orangutan named Marie who had lost both arms early in life. Marie made precisely the kind of connection between her own body and that of another individual required for cognitive empathy. Savage-Rumbaugh, one of whose fingers is missing its tip, was grooming and talking to Marie, when the latter suddenly noticed the stumpy finger. She inspected it closely, holding Savage-Rumbaugh's hand with one of her feet. Marie then brought Sue's finger into contact with the stump of one of her own arms. She looked questioningly into her companion's eyes as if wondering if she saw the same connection.

10. As can be seen from the accompanying figure, Azalea's grooming activity was far below that of her peers until the age of 18 months. She received an average amount of grooming until this age, after which she began to receive substantially more than her peers. It is not clear whether these two developments were related. Data from de Waal et al. (1995).

11. Fedigan and Fedigan, 1977, p. 215.

12. In the sort of experiment that, in the words of Silk (1992a), makes one wonder about taxonomic bounds for compassion, Berkson experimentally blinded a number of young rhesus monkeys in a free-ranging population. These young monkeys, groping for roots as they went, had great difficulty in finding their way through the mangrove. Their mothers stopped often and waited for them, and the group as a whole was extra vigilant and alarmed if the blind infants were approached by human

observers. "The blind babies were never left completely alone. . . . It is remarkable that there was always another animal in the group near them. In addition, two individuals who were unrelated to the mothers often stayed with the blind infants during this time" (Berkson, 1973, p. 585).

13. Scanlon, 1986, p. 107.

14. It is surprising how little we know about precisely which situations trigger succorance and which generate intolerance. Pavelka (1993, p. 92) describing free-ranging Japanese macaques, comments that incapacitated group members may meet with hostility. This was the reaction to staggering and stumbling individuals who were recovering from anesthesia after routine veterinary checkups. Even though the mothers would protect their drowsy infants, they sometimes bit and shook their offspring as if punishing them for inappropriate behavior. Over the years the monkeys became used to the sight of half-sedated group mates, and aggressivity toward them diminished.

15. Moss, 1988, p. 73.

16. Goodall, 1990, p. 196.

17. Flint was perhaps too old to be adopted by others. Younger orphans are frequently taken care of by female relatives, sometimes by unrelated females, and enjoy remarkable tolerance from adult males (Goodall, 1986, pp. 102–103; Nishida, 1979, p. 106). For a review of adoption in nonhuman primates see Thierry and Anderson (1986).

In the Arnhem chimpanzee colony, Fons lost his mother when he was four. Soon afterward he was seen associating with adult males, particularly his presumed father, Luit, who became his mighty protector. Fons resembled Luit even at an early age (de Waal, 1982, p. 75), but he looks, acts, and sounds uncannily like the older male now that he is fully grown. Unfortunately, Luit died before the colony could be subjected to a paternity analysis.

18. Smuts, 1985, p. 23.

19. No agreed-upon explanation for human weeping exists. Outside the primate order, copious tear production is associated with marine habitats (probably because of the increased need for salt excretion). Thus, fresh-water crocodiles do not shed tears, whereas sea crocodiles do. Tears can also be observed in sea otters and seals. Because of this link with marine ecology, proponents of the "aquatic ape" theory view human tears as evidence that there must have been an aquatic phase during human evolution (Morgan, 1982).

Accounts of tears in primates other than humans are extremely rare, and need to be considered with reservation. Most likely, people *expect* tears, hence project or imagine them. The only report by an experienced primatologist is in Fossey (1983). Despite my profound skepticism— could the "tears" have been due to excessive perspiration?—I present it here because Fossey was certainly aware of the extraordinary nature of her observation.

The account concerns Coco, a young mountain gorilla whose entire family had been wiped out by poachers. Following weeks of mistreatment and life in a tiny box, the young gorilla was claimed by Fossey and brought to her camp. There Coco for the first time saw her natural environment again. "Coco sat on my lap calmly for a few minutes before walking to a long bench below the window that overlooked nearby slopes of Visoke. With great difficulty she climbed onto the bench and gazed out at the mountain. Suddenly she began to sob and shed actual tears, something I have never seen a gorilla do before or since. When it finally grew dark she curled up in a nest of vegetation I had made for her and softly whimpered herself to sleep" (p. 110).

20. Yerkes and Yerkes, 1929, p. 297.

21. Temerlin, 1975, p. 165.

22. Boesch, 1992, p. 149.

23. The German naturalist Bernhard Grzimek was once attacked by an adult male chimpanzee, an event he was lucky to survive. When his rage had died down, the ape seemed very concerned about the outcome. He approached the professor and tried, with his fingers, to close and press together the edges of the worst wounds. Lorenz (1967, p. 215), who described this incident, adds that "it is highly characteristic of that dauntless scientist that he permitted the ape to do so."

24. The accompanying figure shows data from de Waal and Aureli (forthcoming) concerning 1,321 spontaneous aggressive incidents in an outdoor colony of twenty chimpanzees at the Yerkes Field Station. The graph demonstrates that immediately after fights (in the first two minutes) bystanders often make contact with participants in the conflict, particularly with the recipients of serious aggression.

25. Skinner (1990) saw both cognitive psychology and creationism as heavily influenced by religion. This element is obvious enough with regard to creationism, sometimes mislabeled creation science ("mislabeled" because creationists work with a single hypothesis, determined a priori to be true, whereas science tries to choose among alternative hypotheses). The effect of religion on cognitive psychology may be less evident—hidden as it is by centuries of sophisticated philosophizing, it is revealed in the persistent mind/body and human/animal dualisms. These dualisms lack a factual basis, and psychology would be much better off without them (Gibson, 1994).

26. Most readers would long ago have laid aside this book if I had limited myself to purely descriptive, technical language. There is a fine but important line between the use of anthropomorphism for communicatory purposes or as a heuristic device, and gratuitous anthropomorphism that projects human emotions and intentions onto animals without

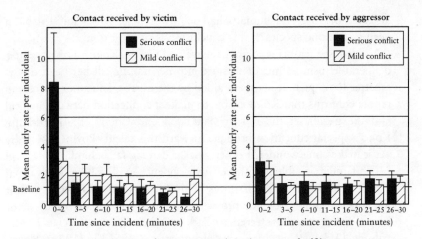

Mean (plus standard error of mean, SEM) hourly rates of affiliative contact (kissing, hugging, grooming, gentle touching) received from bystanders by individuals recently involved in fights. Contacts between former opponents (reconciliations) are excluded from the analysis. The thirty-minute window following each incident has been divided into blocks of two, three, and five minutes. Data are presented separately for incidents involving vocalizations and/or physical contact (serious conflicts) and mere silent threats and lunges (mild conflicts). The baseline shows the mean hourly contact rate received per individual.

justification, explication, or critical investigation. Strong opinions about the use and abuse of anthropomorphism can be found in Kennedy (1992), Marshall Thomas (1993), Masson and McCarthy (1995), and in Mitchell, Thompson, and Miles (forthcoming).

27. Burghardt, 1985, p. 917.

28. Diogenes Laertius, quoted in Menzel, 1986, p. 167.

29. Promises and problems of the cognitive approach to animal behavior have been extensively debated among ethologists. See Griffin (1976, 1984), Kummer (1982), Kummer, Dasser, and Hoyningen-Huene (1990), de Waal (1982, 1991a), and Cheney and Seyfarth (1990).

30. Carpenter, 1975, pp. 452–453.

31. This passage was translated from the Dutch by Kortlandt (1991, p. 11). Anton Portielje was a remarkable observer: he was also the first to notice enough difference between a chimpanzee and a bonobo to suggest, in 1916, that they might be different species. The distinction became official only in 1929 (de Waal, 1989, pp. 177–178).

32. Oddly enough, gorillas were long assumed incapable of passing Gallup's mirror-recognition test (reviewed by Povinelli, 1987). Since Westergaard, Hyatt, and Hopkins (1994) found that bonobos recognize themselves, the gorilla would have been the only anthropoid ape without self-recognition. Apart from not making much evolutionary sense (if the common

ancestor of apes and humans had self-awareness, why should it have been lost in one species?), this conclusion is open to doubt.

A videotape featuring a language-trained gorilla, Koko, shows the ape deliberately using a mirror to stare into her mouth, tilt her head to get a better look, pick at her teeth aided by the mirror, and so on. A recent report confirms that Koko is able to make a connection between herself and the gorilla in the mirror (Patterson and Cohn, 1994). Perhaps Koko's special education brought forward this talent (Povinelli, 1994), yet it makes one wonder if other gorillas can lag far behind.

The familiar problem with negative evidence applies here. Even if we accept that passing the mirror-test demonstrates self-awareness, failing the test certainly does not prove its absence. For debate about these issues, the reader is referred to Parker, Mitchell, and Boccia (1994), volume 11(3) of *New Ideas in Psychology*, pp. 295–377 (1993), Heyes (1993), and Cenami Spada et al. (forthcoming). An experiment promising to turn the presumed gap between monkeys and apes into a gray zone was presented recently by Howell, Kinsey, and Novak (1994).

33. Hatfield, Cacioppo, and Rapson (1993, p. 96) define emotional contagion as "the tendency to automatically mimic and synchronize expressions, vocalizations, postures, and movements with those of another person and, consequently, to converge emotionally."

34. This correlation was reported by Johnson (1982) and Bischof-Köhler (1988). According to the latter study, the link between mirror self-recognition and the emergence of cognitive empathy holds up even after correction for age.

35. The main alternative to a cognitive explanation of the absence of consolation in macaques is the so-called social constraints hypothesis. It posits that macaques run serious risks in associating with an individual who has just been attacked. With their more tolerant and flexible relationships, chimpanzees may not be operating under the same constraints. We plan to conduct experiments to eliminate the risk of approaching a victim of aggression. If macaques still fail to contact distressed group members under these circumstances, the social constraints hypothesis would be weakened (de Waal and Aureli, forthcoming).

36. Certain birds weave leaves into their feathers, and hermit crabs tote around entire houses, replacing them as they grow with larger residences. These self-enhancements are identical in all members of the species and probably have nothing to do with self-awareness. The only nonprimates in which self-decoration might accompany awareness of its effect on others are dolphins and killer whales. Marine mammal trainers speak of their subjects' *jewelry,* meaning the bits of seaweed they string around their pectoral fins or flukes, or the dead fish they carry on their

snouts (Pryor, 1975). Wild dolphins, too, tend to drag "stuff" around (Christine Johnson, personal communication).

If this is indeed self-decoration and not mere object play, it is intriguing in view of the highly developed succorant tendencies in the same mammals. Could the Cetacea, which after all have unusually large brains, be another group with increased awareness of the self? See some fascinating speculations by the neuroanatomist Harry Jerison (1986), and the first studies of dolphin self-recognition by Marten and Psarakos (1994) and Marino, Reiss, and Gallup (1994).

37. Even if apes are the most conspicuous behavioral copiers, they may not be the only ones. An intriguing monkey example comes from Breuggeman (1973, p. 196), who saw a juvenile rhesus monkey follow her mother while the mother carried a newborn. The daughter picked up a piece of coconut shell, carrying it ventrally in the same way that her mother held her new brother. When the mother lay down on her side, with one hand resting on the infant's back, the daugher did the same a few feet away, adopting the exact same posture while holding the shell.

38. Garner, 1896, p. 91.

39. Renewed experimentation on observational learning in monkeys and apes was pioneered by an Italian ethologist, Elisabetta Visalberghi, and an American developmental psychologist, Michael Tomasello. Thus far their findings have failed to support claims of full-blown imitation in nonhuman primates.

Field-workers, such as Boesch (1991a, 1993) and McGrew (1992, pp. 82–87), are not convinced that absence of imitation in the laboratory implies absence in the natural habitat. Although not saying in so many words that captive chimpanzees are backward, they imply it when pointing out that one cannot expect individuals under impoverished conditions to be competent to perform complex tasks. This hypothesis is contradicted by the masterly tool use for which captive orangutans are known, whereas their wild counterparts rarely demonstrate anything close to this ability (Lethmate, 1977; McGrew, 1992). Furthermore, it has been argued that free time under captive conditions actually *promotes* innovation and social sophistication (Kummer and Goodall, 1985).

Clearly, the real issue is not whether captive or wild primates are smarter, but whether the relevant variables have been controlled. The laboratory wins hands down on this count: observational learning covers a wide range of processes that cannot easily be disentangled in the field. For further discussion see Galef (1988), Visalberghi and Fragaszy (1990), Whiten and Ham (1992), Tomasello, Kruger, and Ratner (1993), and Byrne (1995).

40. Menzel, 1974, pp. 134–135.

41. The ability to attribute knowledge, feelings, and intentions to others is now often phrased as the possession of a "theory of mind" about others. This expression derives from an experiment by Premack and Woodruff (1978) in which apes were challenged to infer the intentions of other individuals by watching their efforts on video. The apes seemed to have an idea of the mental states of others. Theory-of-mind research covers both child and nonhuman primate behavior (reviewed in Buttersworth et al., 1991; Whiten, 1991; Byrne, 1995).

According to Cheney and Seyfarth (1991, p. 253), even the most compelling examples of attribution in monkeys and apes "can usually be explained in terms of learned behavioral contingencies, without recourse to higher-order intentionality. What little evidence there is suggests that apes, in particular, may have a theory of mind, but not one that allows them to differentiate clearly or easily among different theories or different minds." This passage was written, however, before Povinelli's experiments strengthened the case for attribution and perspective-taking in the chimpanzee (Povinelli et al., 1990, 1992).

A serious problem with studies of the ape's theory of mind is the interpretation of negative results. The experimental subjects are sometimes presented with rather unusual situations, such as blindfolded persons or persons instructed to stare into the distance. Like us, apes are very sensitive to body language: an unresponsive human experimenter is likely to confuse and disturb them. In addition, the rules of eye contact are different in apes than in humans: rather than gazing directly at others—which they do under exceptional circumstances only, such as during a reconciliation—apes are masters at monitoring companions by means of peripheral vision and quick glances that are barely noticeable. Negative test results may therefore say more about the apes' expectations about normal social interaction and the species barrier in this regard than about their grasp of the connection between looking and seeing.

The fairest comparison of apes tested by humans would be human children tested by apes; who knows how poorly children would do under such circumstances.

42. Example 1: de Waal, 1986b, p. 233; Example 2: de Waal, 1992d, p. 86; Example 3: de Waal, 1986b, p. 238; Example 4: de Waal, 1982, p. 49.
43. Menzel, 1988, p. 258.
44. Salk (1973) discovered the left-side cradling preference of human mothers, and Manning, Heaton, and Chamberlain (1994) report the same bias in gorillas and chimpanzees but not orangutans. See Hopkins and Morris (1993) for a review.
45. Mercer, 1972, p. 123.
46. Goodall, 1971, p. 221.
47. One notably different reaction to polio victims involved two adult males

suspected to be siblings or nephews: Mr. McGregor (with paralyzed legs) and Humphrey (unafflicted). Humphrey stood by McGregor until the end, defending him against even the most dominant aggressors. After McGregor's death, Humphrey kept returning for nearly six months to the place where his possible relative had spent the last days of his life in great pain and misery (Goodall, 1971, pp. 222–224).

48. Nevertheless, members of one species are sometimes vicariously aroused by those of another species. Recently the entire chimpanzee colony at the Yerkes field station was intently following how animal care staff caught an escaped rhesus monkey in the forest around their enclosure. Attempts to lure the monkey back into his cage had failed. The situation became hairy when he climbed a tree. I heard one of the watching chimpanzees, a juvenile named Bjorn, suddenly utter whimpers while seeking reassurance from an ape next to him, reaching out to her. When I looked up, I saw that Bjorn's distressed reaction coincided with the monkey's clinging desperately to a lower branch of the tree; he had just been shot with a tranquilizer dart. People were waiting beneath the tree with a net. Although it was not a situation Bjorn himself had ever been in, he appeared to empathize with the monkey: he uttered another whimper when the escapee dropped into the net.

49. Turnbull, 1972, pp. 112, 230.

50. It is hard to imagine being delighted by the misery of others unless one has a bone to pick with them. Turnbull's (1972) observations have not gone unchallenged; one fascinating speculation has been that the anthropologist felt so isolated and frustrated living with the Ik that *he himself* began to derive joy from their misfortune (Heine, 1985).

51. Weiss et al., 1971, p. 1263.

52. In a careful review of the psychological literature, Batson (1990) compares the attitude of science toward human altruism to that of the Victorians toward sex: it is denied and explained away. All too often, caring for others is interpreted as caring about oneself.

Experiments have failed to confirm this interpretation. Because there is no evidence for selfish motives behind *all* helping behavior, Batson concludes that people do possess a genuine caring capacity. Wispé (1991) supports this view, arguing that feelings of sympathy evaporate the moment self-interest enters as a conscious motive. This is not to deny an internal reward, but the reward seems specifically tied to the other's well-being. Alleviating another person's pain or burden gives a special satisfaction that is simply unavailable to those who base their help on hope for return favors, a desire for praise, or a wish to go to heaven. Such calculations may *mix* with sympathy, but they cannot replace it because "rewards are what sympathy is *not* about . . . Even if one always derived pleasure from helping others, it would not follow that one helps others in order to feel pleasure" (Wispé, 1991, p. 81).

Note that the issue of unselfishness is treated here from the perspective of motivation and conscious intent. Nothing is said about the possibility that acts of sympathy and cognitive altruism may, in the long run and perhaps quite circuitously, serve the actor's self-interest. Indeed, without such benefits the entire complex of empathy and helping behavior could never have evolved. The main point is that these benefits need not factor into the actor's conscious decision-making.

53. Wilson, 1993, p. 50.

3. Rank and Order

1. Hall, 1964, p. 56.
2. Example 1: author's translation of Trumler, 1974, pp. 52–53; Example 2: Lopez, 1978, p. 33; Example 3: Barbara Smuts, personal communication; Example 4: von Stephanitz, 1950, p. 814.
3. Modified from de Waal, 1991b, p. 336.
4. See de Waal (1982, p. 207). Nishida (1994, pp. 390–391) reports similar outraged reactions to violations of the social code. For example, he once saw a wild chimpanzee attack another from behind after a stealthy approach. This tactic is highly unusual: chimpanzees normally signal aggressive intentions in advance. Loudly screaming, the victim chased the attacker, who, although dominant, kept his distance. Speculating that the dominant did not fight back because he felt guilty about the sneak attack, Nishida concludes that "unusually fierce, prolonged retaliation on the part of a subordinate party and the corresponding reluctance to escalate the fight on the part of the dominant party may be one of the factors discriminating moralistic aggression from a conventional counterattack by a subdominant." This characterization of moralistic aggression may also apply to the extraordinary retaliation by Shade, a Japanese macaque, recounted in Chapter 4.
5. Hobbes, 1991 (1651), p. 70.
6. For decades the received view has been that animals engage in approach-retreat encounters, fights, and competitions that may *reveal* who dominates whom, but that the rank order itself is a mere construct of the human observer: animals neither classify themselves in terms of who dominates whom nor do they deliberately strive for better positions (Bernstein, 1981; Altmann, 1981; Mason, 1993).

Less prominent in the textbooks, yet present at least since Maslow (1936), is the alternative that social dominance does exist in the minds of animals. For example, research on chimpanzees has produced evidence supporting what appear to be calculated, Machiavellian strategies to attain high status (de Waal, 1982). Theory formation tends to follow a pendulum pattern, and future students of animal behavior will no doubt revisit the issue of status striving. It is unclear at this point, for instance, how much the "striving" position really differs from the following formulation of the "nonstriving" position: "The critical issue is not social status per se, but the extent to which the relationship is oppressive and the type of satisfactions it affords. When the oppressive constraints in a relationship consistently exceed the satisfactions for one of the participants, the potential for conflict is high. . . . Primates generally act so as to maximize their personal freedom and mobility under demanding circumstances" (Mason, 1993, p. 25).

7. Whereas the outcome of conflict varies with the presence or absence of allies, or the resource at stake, expressions of submission are virtually immune to such effects. The bared-teeth face of some macaque species or the chimpanzee's bobbing and pant-grunting are completely predictable: it is hard even to imagine a situation in which a dominant would give these signals to a subordinate (Noë, de Waal, and van Hooff, 1980; de Waal and Luttrell, 1985). The simplest explanation for the independence of context is a cognitive one, namely that primates know in which relationship they are dominant or subordinate and that they communicate these evaluations to one another. In this view, submissive gestures and facial expressions reveal how the animals *themselves* perceive dominance relationships.

8. My text speaks of male, not female, rivals because it is written with an eye on chimpanzees, a species in which males are by far the more hierarchical sex. Most male animals compete about mates; females compete over food for themselves and their offspring. Whereas male chimpanzees are no exception to this rule, female chimpanzees at the best-known field sites avoid competition by living dispersed throughout the forest, each female occupying her own core area (Wrangham, 1979). This tendency may explain the lesser development of the female hierarchy in these apes. Most monkeys, on the other hand, form permanent mixed-sex groups in which food competition is common: females of these species are as dominance oriented as males.

   Accounts of status competition in both male and female monkeys may be found in Bernstein (1969), Chance, Emory, and Payne (1977), de Waal (1977; 1989a, pp. 133–140), Leonard (1979), Walters (1980), and Small (1990).

9. Koestler, quoted in Barlow, 1991, p. 91.

10. The effect of submission on social relationships was investigated during three months of instability between two male chimpanzees, Yeroen and Luit, at the Arnhem Zoo (de Waal, 1986c). Over two hundred hours of observations were collected in order to document changes around the time of first submission. During the weeks prior to this moment, friendly contacts became increasingly rare, whereas aggressive confrontations and intimidation displays between the two males reached their peak. As usual during power struggles, the exchange of status signals ceased

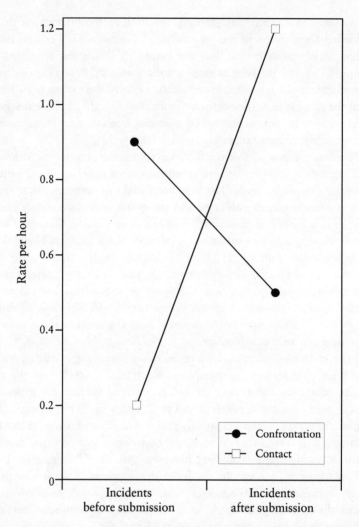

On March 16, 1978, Yeroen bowed and pant-grunted to Luit for the first time after three months of intense rivalry. Shown are mean hourly rates of aggressive confrontations and friendly contacts between the two males during seven observation days prior to and following Yeroen's first submission.

completely. The low level of contact occurred because the future dominant, Luit, walked away each time Yeroen approached.

Luit's attitude changed dramatically the day Yeroen uttered his first submissive pant-grunts: Luit suddenly became receptive to Yeroen's overtures. Aggression diminished and contact between the rivals increased over the next few days. Significantly, the longest grooming session of the entire period took place only a few hours after Yeroen's acknowledgment of Luit's status. This case study confirmed that peaceful coexistence among male chimpanzees depends on a formal clarification of their dominance relationship.

11. Dog owners who believe in the efficacy of punishment after the fact may be right for the wrong reasons. Instead of having taught their pet to associate a bad habit with punishment, they may have taught them merely to avoid certain sights, smells, or locations associated with the transgression. For example, a dog learns to avoid a bedroom in which he was admonished for chewing a pair of shoes, hence stops the chewing inasmuch as he does not encounter the same shoes again (Vollmer, 1977).

12. The overly submissive macaques illustrate the advantage of attributional capacities found in humans, and perhaps in apes (Chapter 2). A chimpanzee confronted with similar behavior in a subordinate would probably suspect that something had been going on, in the same way that we guess from the absence of a happy greeting by a dog that a rule must have been broken. Because of this more complete understanding of what may lie behind an expression of emotion, chimpanzees and people need to be more careful: better lie-detecting abilities require more sophisticated lies. I have never seen subordinate male chimpanzees react in the same way as these macaques. Even when caught in the middle of a prohibited act, they tend first to try concealment of the evidence (such as dropping their hands over their penis) before responding with fear and submission.

13. Coe and Rosenblum, 1984, p. 51.

14. Davis, 1989, p. 88.

15. Scott, 1971, p. 81.

16. Darwin, 1965 (1872), p. 309.

17. Starting with Trivers' (1971) theory of reciprocal altruism, we now seem to have the first elements of a plausible scenario for the evolution of conscience. My discussion closely follows recent theorizing by Alexander (1987), Frank (1988), and Simon (1990) about susceptibility to social influence (Simon's *docility*) and emotional commitment (Frank). A central concept is reputation, or status, as defined by Alexander (1987, p. 95): "Systems of indirect reciprocity, and therefore moral systems, are social systems structured around the importance of status. The concept of status simply implies that an individual's privileges, or its access to resources, are controlled in part by how others collectively think of him as a result of past interactions."

18. A sharp distinction between the reason for the evolution of a behavior and its underlying motivation is assumed in discussions of evolutionary ethics by Frank (1988), Simon (1990), and myself (Chapter 1).

Alexander (1987, 1993), in contrast, defends the position that, consciously or not, selfish motives guide all human behavior, including behavior that the actors themselves believe to be unselfish. Similarly,

Badcock (1986) tries to reconcile our species' true nature—which he sees as literally selfish—with altruistic motives. To the rescue comes Freudian repression theory: the ego defends itself by presenting a distorted picture of internal processes. It hides selfish motives from our conscious self, presenting them in the most favorable light. In the view of Alexander and Badcock, then, the human mind expends a good deal of energy in concealing selfish agendas: we are sophisticated hypocrites.

In a review of Wright (1994), who adopts the same position, Steven Pinker (1994, p. 35), a cognitive neuroscientist, makes the by-now-familiar point (Chapter 1 notes 12 and 27) that this view confuses proximate and ultimate causation: "When a person's public stance is selfless but his private motives serve his interests, we can call it hypocrisy. However, when a person's public stance and private motives are both selfless but those motives came about because they once served the interests of his ancestors' genes, we have not uncovered hypocrisy; we have invoked a scientific explanation couched at a different level of analysis. . . . The evolutionary *causes* of our motives can't be judged as if they *are* our motives."

19. Lever, 1976, pp. 482, 483.
20. Gilligan, 1982, p. 104.
21. Gilligan has been accused of giving in to popular expectations about gender roles without careful research to back her position (Colby and Damon, 1983; Walker, 1984; Mednick, 1989; Smetana, Killen, and Turiel, 1991). Some believe that her ideas stand in the way of progress, a criticism leveled most harshly by Broughton (1983, p. 614): "Gilligan does not seem very concerned with societal transformation given her desire to imbed women even more deeply in the domestic and personal aspects of welfare in civil society."
22. Walker, 1984, p. 687.
23. Hoffman, 1978, p. 718.
24. Applying ethological methods of data collection in the field, Edwards (1993) found consistent sex differences across a dozen cultures in countries from Kenya to India, and from the Philippines to the United States. During middle childhood, boys spend more time away from the home than girls, and girls have more contact with and responsibility for infants. The latter difference probably reflects both girls' greater attraction to infants and socialization practices (mothers preferentially assigning child care to daughters).
25. Strier, 1992, p. 85.
26. The Finnish study is special in that it did not focus on open conflict only—which universally is most common in boys (Maccoby and Jacklin, 1974)—but included *indirect* conflict by asking children what they do when angry. A rather cruel picture of girls emerged: they said they could

stay mad forever, whereas boys measured their anger in minutes (Lager-spetz, Björkqvist, and Peltonen, 1988).

A work of fiction detailing female-style conflict is Margaret Atwood's *Cat's Eye.* The author contrasts the torments to which girls subject one another with the more straightforward competition among boys. At one point, the principal female character complains: "I considered telling my [elder] brother, asking him for help. But tell what exactly? I have no black eyes, no bloody noses to report: Cordelia does nothing physical. If it was boys, chasing or teasing, he would know what to do, but I don't suffer from boys in this way. Against girls and their indirectness, their whisperings, he would be helpless" (Atwood, 1989, p. 166).

Unfortunately, sex differences in the domain of interpersonal conflict have been only poorly investigated. We are in the curious situation of knowing more about spontaneous aggression and reconciliation in non-human primates than in our own species (de Waal, 1989a).

27. Tannen, 1990, p. 150.

28. In well-established chimpanzee societies, males reconcile more readily after fights than females, and the male power structure is well defined and fiercely contested compared to the rather informal female hierarchy (Goodall, 1971, 1986; Bygott, 1979; de Waal, 1982, 1986c). As explained by Nishida (1989, p. 86), females decide dominance issues on the basis of seniority: "Unlike males, whose reproductive success depends on social status, female reproductive success may depend primarily upon acquiring a core area near the center of the unit-group's territory. Therefore, females who have acquired their own core areas have no pressing reason to strive for higher rank. Thus, a female's rank will be more or less fixed sometime after her immigration. Thereafter, her promotion in rank will be caused mainly by the death of senior high-ranking females and by the addition of younger low-ranking females to the hierarchy."

The rarity of overt conflict between female chimpanzees, and their relatively vague hierarchy, should not be interpreted as a lack of interest in competitive goals. Under particular circumstances, such as when wild females migrate into another community or when captive females are introduced to strangers, females do compete fiercely. In Arnhem, females did not give up their top positions willingly when adult males were introduced into the colony (de Waal, 1982); and when a new colony was established at the Detroit Zoo, females engaged in dominance strategies remarkably similar to those of males (Baker and Smuts, 1994).

Given the plasticity of the behavior of chimpanzees and other primates, researchers have moved away from deterministic explanations of sex differences. Discussions of these differences increasingly pay attention to potential, context dependency, social values, and payoff curves

(Goldfoot and Neff, 1985; Smuts, 1987; de Waal, 1993b; Baker and Smuts, 1994).

29. The effect of social context on the nurturing abilities of rhesus monkeys was investigated by Judith Gibber and Robert Goy (Gibber, 1981, pp. 63–66). In one series of tests, individually-caged monkeys received an unfamiliar infant. Rhesus males normally pay little attention to young infants, but under these conditions they proved remarkably friendly. This response was not seen in the second test series, however, which involved pair-housed subjects. Males who individually had picked up and held infants failed to do so in the presence of a female companion: the female did most of the infant handling. Males apparently defer to the other sex when it comes to infant care.

30. Andries Vierlingh, translated by Schama, 1987, p. 43.

31. See Schama (1987, pp. 25–50) for a remarkable outsider account of Dutch history. That the tendencies noted are still part of the culture was demonstrated in 1995, when river levels in the Netherlands rose to the point that thousands of families had to be moved to higher land. In a show of solidarity, Queen Beatrix appeared in rubber boots at the threatened dikes.

32. These socioecological theories were developed by, among others, Wrangham (1980), Vehrencamp (1983), and van Schaik (1989). Another important variable, emphasized by van Schaik, is within-group competition. The energy invested in rank-related affairs obviously depends on the advantage of high rank. This advantage varies with the kind of food a species lives on: there is little point in trying to monopolize scattered low-energy foods, such as foliage; but it does pay to be dominant in the case of clumped high-energy foods such as fruits. A steep dominance hierarchy is therefore more likely under the latter condition. For a discussion of contrasting dominance styles in monkeys and apes see de Waal (1989b) and de Waal and Luttrell (1989).

The idea that despotic dominance requires a closed exit door most likely applies to humans as well. For example, domestic abuse may persist especially when a victim's (perceived) options to quit the relationship are minimal. Conversely, a totalitarian regime may lose its grip once the national border has become porous. This happened when changes in neighboring countries created an opening for an exodus from East Germany. Erich Honecker's power evaporated as soon as people began voting with their feet.

33. For anthropological and evolutionary perspectives on despotism and egalitarianism see Woodburn (1982), Betzig (1986), Knauft (1991), Boehm (1993), and Erdal and Whiten (1994, in press). Boehm was the first to fully develop the idea that egalitarianism is not simply the absence of social stratification but the product of vigilance against excessive individual ambition.

If hierarchical tendencies are counteracted in egalitarian societies, it is precisely because they have not disappeared. Thus, rather than fitting a naive, idealized picture of human nature, egalitarianism occurs in full recognition of the ubiquitous tendency of men to accumulate power and privilege. There is only one way to neutralize this tendency: alliances from below. Ridicule and social control are important in holding ambition in check, but cannot work without sanctions. And sanctions against the top ultimately require joint action by lower levels.

Alliances from below are also recognizable in the balance-of-power arrangements of chimpanzees (de Waal, 1982, 1984). Most likely, therefore, the common ancestor of humans and apes already had a dominance orientation *and* leveling tendencies. Knauft (1991) is no doubt correct, however, that humans took the leveling tendency a giant step further by means of cultural norms and institutions.

Inasmuch as democracy can be interpreted as a hierarchical arrangement achieved by egalitarian means, the evolution and history of leveling mechanisms is relevant in relation not only to small-scale human societies but also to state organization.

34. Erdal and Whiten, in press.
35. De Waal, 1982, p. 124.
36. Usually, the control role in primate groups is restricted to a single dominant male. Although female macaques without relatives have been known to perform control activities (Varley and Symmes, 1966; Reinhardt, Dodsworth, and Scanlan, 1986), and a female chimpanzee may stop a fight if she is the highest-ranking individual present (de Waal, 1982; Boehm, 1992), the strong commitment of female primates to close relatives—and perhaps by extension to close friends—hampers impartial arbitration.

In the early years, the Arnhem chimpanzee colony was dominated by Mama, a female without offspring. Unfortunately, we have little information on her rule. We do know from veterinary records that after males took over, the number of serious injuries in the colony dropped sharply. One possible explanation is that males were more effective than Mama at controlling fights (de Waal, 1982).

In a cohousing experiment at the Wisconsin Primate Center with juvenile rhesus and stump-tail monkeys (Chapter 5), one mixed-species group was dominated by a male stump-tail, the other by a female stump-tail. Only the male performed control interventions, a task overriding the typical own-species bias. He intervened far more often than any other monkey, usually protecting losers even if it meant favoring rhesus monkeys over stump-tails (as he did in 67 percent of forty-six interventions). All other stump-tails favored their own species. The dominant female in the other group intervened only seven times, five of which were in support of other stump-tails.

For discussion of the control role see de Waal (1977, 1984), Ehardt and Bernstein (1992), and Boehm (1992, 1994).

37. Boehm, 1992, p. 147.

38. When Goblin, after having lost his top position in the Gombe community, tried to stage a comeback, he was defeated by a ferocious mass attack. The unusually hostile reception may have had something to do with the fact that Goblin had been a very tempestuous alpha male who frequently disrupted the group with his charging displays. "Possibly, his return would have roused a less dramatic response had he himself been a more peaceful and calm individual" (Goodall, 1992, p. 139).

39. The role of external threats, particularly from enemy groups of the same species, in the evolution of moral systems is a recurrent theme treated also in Chapters 1 and 5, and emphasized by both Darwin (1871) and Alexander (1987).

4. Quid pro Quo

1. Isaac, 1978, p. 107.

2. This rule cannot explain how altruistic exchanges started. A mere response tendency ("Do as the other did") does not suffice. According to computer simulations of tit-for-tat strategies, an initial cooperative attitude is necessary (Axelrod and Hamilton, 1981).

3. Indeed, this attitude is so unhumanlike that it may develop only under the most extreme circumstances. See Chapter 2 for Turnbull's (1972) claim that the Ik abandoned morality in the face of severe food shortages.

4. Milton, 1992, p. 39.

5. These ideas have been summarized by another anthropologist, Kristen Hawkes, who argues (1990) that men have an interest in providing food bonanzas that can feed many hungry mouths. Hunting success and generosity increase a male's attractiveness as a mate and help foster political ties. The meat distribution strategies of male chimpanzees discussed later on in this chapter seem entirely compatible with Hawkes's "showing off" hypothesis. See also note 20 to this chapter.

6. Lee, 1969, p. 62.

7. According to the French sociologist Claude Fischler, meat blurs the line between ourselves and what we eat. We both challenge and confirm our self-identity when consuming another being made of blood, bones, brains, secretions, and excretions. In this view, human flesh is the superlative meat; and Fischler (1990) indeed argues in *L'Homnivore* (a wordplay making a maneater out of an omnivore) that our obsession with animal foods ultimately derives from ancient practices of cannibalism and human sacrifice. A link with this past is also preserved in the Roman

Catholic ritual of ingesting bread and wine as representations of Christ's flesh and blood.

8. Nishida et al., 1992, p. 169.

9. De Waal, 1982, p. 110.

10. According to Sahlins (1965), human reciprocity takes two forms: (a) "vice versa" movements of goods and services within dyadic relationships, and (b) centralized exchanges via a recognized authority who pools and redistributes resources. The latter function may be performed by a chief or, in modern society, the government.

    Centralized reciprocity is recognizable in both the food distribution strategies of top-ranking chimpanzees and their control role—their tendency to defend the weak against the strong (Chapter 3). In both cases a dominant individual dampens competition to the advantage of low-ranking members, in return receiving support and respect.

    Studies on aid giving and sharing in human children, too, fit this pattern. Apart from reciprocity at the dyadic level, they demonstrate increasing protectiveness and generosity with increasing status (Ginsburg and Miller, 1981; Birch and Billman, 1986; Grammer, 1992).

11. Detailed information about chimpanzee predation and meat sharing can be found in Teleki (1973b), Goodall (1986), Boesch and Boesch (1989), Boesch (1994ab), and Stanford et al. (1994a).

    Whether bonobos fit the predation hypothesis of the evolution of sharing is still ambiguous. Despite reports of meat sharing (Badrian and Malenky, 1984; Ihobe, 1992; Hohmann and Fruth, 1993), predation seems relatively insignificant. For example, bonobos have not been observed to hunt monkeys; on the contrary, their relations with monkeys seem rather friendly (Ihobe, 1990; Sabater Pi et al., 1993).

    Plant food sharing, in contrast, is common. At a provisioning site for bonobos in Zaire, sugarcane was widely shared (Kuroda, 1984), and bonobos have also been seen to divide large *Treculia* and *Anonidium* fruits, some weighing 30 kilograms apiece (Hohmann and Fruth, forthcoming). Perhaps consumption of these fruits helped promote sharing tendencies. It should be added, though, that my own captive studies suggest that the bonobo's food-related tolerance is no match for that of the chimpanzee (de Waal, 1992b; and note 16 to this chapter).

12. Sharing between mother and offspring occurs in most or all primates, including rhesus monkeys, as does cofeeding on a clumped yet abundant food source. When I speak of "nonsharing" species, I mean primates with a predominantly competitive mode of interaction around food, in which subordinates never remove food directly from the hands or mouth of unrelated dominants. Generally, these species also lack communication signals specific to the sharing context, such as gestures and vocalizations to solicit food.

13. Goodall, 1986, p. 357.

14. D'Amato and Eisenstein, 1972, p. 8.

15. The main effect of food deprivation is to increase interest in food. But the same effect can be achieved by presenting favorite foods. I feel that there is so much to be discovered in this way that food deprivation is wholly unnecessary. The capuchins and chimpanzees in our tests receive foods that they particularly like but do not normally get, at least not in such abundance.

16. Feistner and McGrew (1989), who review primate patterns of food distribution, define sharing as the "transfer of a defensible food-item from one food-motivated individual to another, excluding theft" (p. 22).

    Below is a classification of four methods of interindividual food transfer during interactions relative to plant food in chimpanzees at the Yerkes Primate Center (de Waal, 1989d), bonobos at the San Diego Zoo (de Waal, 1992b), and capuchins at the Wisconsin Primate Center (de Waal, Luttrell, and Canfield, 1993). All methods of transfer occurred in all species, but chimpanzees showed more tolerant transfers (cofeeding and relaxed taking) and bonobos more intolerant transfers (forced claims and theft).

    Forced claim or theft: One individual supplants another at a food source, grabs food by force, or snatches a piece and runs. The first two patterns are typical of high-ranking individuals; the last, of subordinates and juveniles.

    Relaxed taking: One individual, in full view of the possessor, removes food from his or her hands in a relaxed or playful manner without threat signals or use of force.

    Cofeeding: An individual joins the possessor to feed peacefully on the same source, which both may hold. This category includes active food donations: 0.2 percent of all transfers in chimpanzees, 2.7 percent in bonobos, and 1.8 percent in capuchins.

    Nearby collection: An individual waits for dropped pieces and scraps, which are collected from within arm's reach of the possessor.

	Chimpanzees	Bonobos	Capuchins
Number of transfers	2,377	598	931
Forced claim or theft	9.5%	44.5%	26.2%
Relaxed taking	37.1%	15.7%	26.5%
Cofeeding	35.9%	17.6%	9.2%
Nearby collection	17.6%	22.2%	38.0%

17. The effect of celebration was investigated through a variation in the method with which bundles of foliage were provided to the chimpanzees:

Keeper delivery: A caretaker brought bundles from a distance, giving the colony one to two minutes for celebration before the bundles were thrown into the enclosure.

Self-delivery: Well before the trial, I concealed the bundles behind me on the observation platform. At an unexpected moment, I threw them into the enclosure, giving the colony no chance to prepare for the event.

Aggression at the moment that the food arrived was rare under either condition; most of it occurred when individuals tried to join existing feeding clusters, or begged from possessors. On average, self-delivery trials involved 1.5 times more aggressive incidents than keeper-delivery trials. This finding is consistent with the idea that celebration increases social tolerance. The ambiance seemed less explosive during sessions preceded by this flurry of body contact and rituals confirming the hierarchy (de Waal, 1992b).

18. Respect for possession was first investigated in relation to sexual partners in hamadryas baboons. Males of this species do not interfere with each other's bonds with females: even large, totally dominant males are inhibited from taking over the female of another male after having seen the two together for a couple of minutes (Kummer, Götz, and Angst, 1974). Recently, Kummer and his coworkers conducted further experiments on possession and property rules, which showed similar inhibitions in relation to objects (Sigg and Falett, 1985; Kummer and Cords, 1991; Kummer, 1991).

19. The accompanying graph (p. 246) illustrates that shared amounts are correlated and well balanced within pairs of dominant and subordinate chimpanzees. Since a few rank relationships were ambiguous, only thirty of the thirty-six pairings among the nine adults are included. Reciprocity, tested on the entire 9 x 9 matrix, resulted in a correlation of $r = 0.55$ ($P = 0.001$). See de Waal (1989d) for details.

20. Sex-for-food transactions have been documented by Kuroda (1984) for wild bonobos, and by de Waal (1987, 1989a) for captive bonobos. Such exchanges are less prominent but not absent in chimpanzees (Yerkes, 1941; Teleki, 1973b). Thus, the best predictor of hunting by male chimpanzees in Gombe National Park is the presence of estrous females in their traveling party. One motivation for hunting, then, may be to increase mating success through sharing with females (Stanford et al., 1994b).

21. Smuts, 1985, p. 223.

22. In our own economy, the value of various services is expressed in a common currency: money. The evolutionary equivalent of money, reproductive fitness, is hard to measure. A discussion of this issue by Seyfarth and Cheney, with reference to their own field data, is contained in a collection of papers on the current state of knowledge of reciprocal

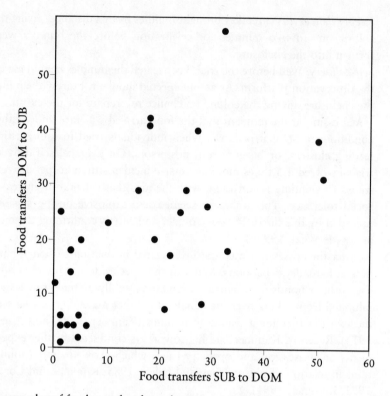

Scatter plot of food transfers from dominant to subordinate chimpanzees (DOM to SUB) against transfers in the opposite direction (SUB to DOM). Each dot represents one adult-adult partner combination.

altruism. See *Ethology and Sociobiology 9* (1988, pp. 67–257). Reciprocity is also a central theme in Harcourt and de Waal's volume on alliance formation (1992).

23. For details see de Waal and Luttrell (1988). The same statistical elimination procedure was applied to food sharing, with identical outcome: a significant level of reciprocity remained after controlling for symmetrical relationship characteristics, such as kinship and association time (de Waal, 1989d).

24. Silk (1992b) challenges the claim that only chimpanzees have a system of direct revenge. The author reports reciprocal *contra* interventions among male bonnet monkeys. Because interventions against the established order were as uncommon in these monkeys as in other macaques, it remains to be seen how comparable their behavior is to that of chimpanzees, who retaliate regardless of the hierarchy (de Waal and Luttrell, 1988). For further data on revenge in monkeys and apes see de Waal (1982, 1989d), Aureli (1992), and Cheney and Seyfarth (1986, 1989).

Discussions from an evolutionary perspective have been provided by Trivers (1971) and by Clutton-Brock and Parker (1995).
25. Jacoby, 1983, p. 13.
26. Chagnon, 1988, p. 990.
27. Huxley, 1989 (1894), p. 140.

5. Getting Along

1. Midgley, 1994, p. 119.
2. Lorenz, 1981, p. 45.
3. Ardrey (1967) called our species the Killer Ape, based on Dart's (1959) depiction of *Australopithecus* as a carnivore who loved to devour writhing flesh and slake its thirst with warm blood (a truly imaginative conclusion, given the limits of fossil data). Dart saw the lust to kill as humanity's "mark of Cain," a feature that sets us apart from our anthropoid relatives. This line of thinking led others to seek the origin of war in hunting, and to present aggressiveness as a prerequisite to progress (Cartmill, 1993).

   One of several problems with the Killer Ape myth is that it equates carnivorousness with violence against one's own kind. We need to distinguish aggression definitively from predation, as Lorenz (1966, p. ix) wisely did when he defined aggression as "the fighting instinct in beast and man which is directed against members of the same species." Thus, it is assumed that a leopard killing an antelope is motivated by hunger, whereas a leopard driving another leopard out of his territory is motivated by aggression. For one thing, members of the same species communicate extensively during confrontations. They have ritualizations and inhibitions to prevent bloodshed, whereas predators usually assail their prey without warning signals and without pulling any punches. The difference in form and purpose of the two kinds of attack is obvious if we compare a cat stalking a mouse with one bristling and hissing at a rival.

   In light of Lorenz's emphasis on this crucial distinction, the most preposterous cover under which his *On Aggression* ever appeared (undoubtedly selected for marketing purposes) was one featuring a dramatic painting of a lion sinking teeth and claws into a frightened horse.
4. Notable exceptions are developmental studies by Sackin and Thelen (1984), Hartup et al. (1988), and Killen and Turiel (1991). Consistent with the theoretical framework presented in this chapter, Killen and Turiel note that fights among children are not always destructive; they often have a social orientation. "Children's conflicts are not solely struggles about selfish desires or aggressive impulses. Children are responsive to others and engage in negotiations to resolve their conflicts" (p. 254).
5. Initially, this flexibility baffled field-workers. Did chimpanzees know

stable groups at all? After years of documenting the composition of chimpanzee parties in the Mahale Mountains, Nishida (1968) was the first to crack the puzzle. He discovered that the ever-changing parties belong to an umbrella group, the members of which mix freely among themselves yet never with members of another such group. Nishida called this higher level of organization a "unit-group," yet the literature came to favor Jane Goodall's term "community" (mere "group" would hardly do, given the fission-fusion character of chimpanzee society).

6. The qualifier "best-known wild populations" is necessary, given the remarkable behavioral diversity—sometimes called cultural variation—of wild chimpanzees. Even populations in similar habitats in the same part of Africa differ in communication gestures and tool technology (Nishida, 1987; McGrew, 1992; Wrangham et al., 1994).

Female-female relations are probably the most variable element of chimpanzee social organization (de Waal, 1994; Baker and Smuts, 1994). These relations range from close in all captive colonies known to me, to rather loose in wild populations at the Gombe and Mahale Mountains national parks in Tanzania (Nishida, 1979; Goodall, 1986) as well as in the Kibale Forest in Uganda (Wrangham, Clark, and Is-abirye-Basuta, 1992).

This variability may not represent an absolute difference between captive and wild conditions, however. Female bonding appears to exist in a small chimpanzee population "trapped" by agricultural encroachment in a forest of approximately 6 square kilometers on top of a mountain in Bossou, Guinea. Despite the substantial amount of available space, Sugiyama (1984, 1988) often saw the majority of individuals in this forest travel together in a single party, and measured relatively high rates of female-female grooming. Similarly, female chimpanzees seem more sociable in Taï Forest, Ivory Coast, than at other sites: they frequently associate, develop friendships, share food, and support one another. Boesch (1991b) attributes this sociability to cooperative defense against leopards.

7. Masters, 1984, p. 209.
8. Van Schaik, 1983, p. 138.
9. Boinsky, 1991, p. 187.
10. Norris, 1991, p. 187.
11. Bowlby, 1981, p. 172.
12. Moss, 1988, p. 125.
13. Using portraits of familiar individuals of both species, Matsuzawa (1989, 1990) compared facial discrimination by chimpanzee and human subjects as measured by latency with finding the correct name (the chimpanzees had first learned to match faces with name-keys on a computer keyboard). He found that recognition of human faces is easier for humans than chimpanzees, and recognition of chimpanzee faces easier

for chimpanzees than humans. Similar experiments in which human subjects select among people of different races have documented an own-race bias, that is, we find it easier to differentiate members of one's own race than members of another race (Brigham and Malpass, 1985).

14. The technique of comparing postconflict and matched-control observations, introduced by de Waal and Yoshihara (1983), has been followed in most observational studies in both captivity and the field. Different control procedures have yielded essentially the same results. See de Waal (1993a) for an overview of the methods used in reconciliation research on nonhuman primates.

15. Altmann, 1980, p. 163.

16. The human own-species bias has been declared morally equivalent to sexism and racism by the Australian animal-rights philosopher Peter Singer (1976). The term he uses for this bias is "speciesism." Although Singer traces its origin to the ancient Greeks and the Bible, thus positing that speciesism is a Western peculiarity, own-species bias certainly antedates humanity itself. Virtually all animals, given a choice, will accord their own kind the better treatment, and cross-species exploitation is absolutely rampant in nature (for instance, some ant species keep aphids as livestock, or force other ant species into labor; Holldöbbler and Wilson, 1990).

17. As expected, stump-tails were characterized by high reconciliation rates (see figure, p. 250). Rhesus monkeys not exposed to stump-tails (rhesus controls) had low rates throughout the experiment. Rhesus living with stump-tails (rhesus subjects), in contrast, started out at the same level as the controls, but rose steadily during the cohousing phases (Co-1, 2, and 3), remaining high even after removal of the stump-tails in the postphase (from de Waal and Johanowicz, 1993).

18. This insight is not limited to Japanese teachers. An early American student of child behavior, Helen Dawe (1934, p. 154), noted that "the mother or teacher who continually interferes is depriving the child of excellent opportunities to learn social adjustment." When Peter Verbeek, a graduate student, observed children at an outdoor playground in Florida, he found that approximately one-third of the fights were broken up by teachers. The teachers would tell the children to "make peace," yet only 8 percent of the combatants continued to associate after a forced reconciliation compared to 35 percent after an independently resolved dispute.

19. Gottman, 1994, p. 41.

20. Kurland (1977) was the first to report high rates of aggression between kin in free-ranging Japanese macaques. This paradoxical observation was confirmed for rhesus and stump-tail monkeys in outdoor enclosures by Bernstein and coworkers (Bernstein and Ehardt, 1986; Bernstein, Judge, and Ruehlmann, 1993) and de Waal and Luttrell (1989). Further-

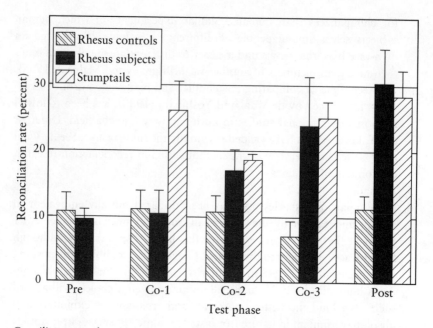

Conciliatory tendency increases in rhesus monkeys housed with stump-tails. Shown are individual means (plus standard error of means, SEM) of aggressive conflicts followed within three minutes by reconciliation. Before and after the experiment (pre- and post-phases), subjects were housed with members of their own species only (no pre-phase data are available on stump-tailed monkeys).

more, kinship ties are not the only bonds characterized by frequent aggression; the same applies to ties among nonkin (de Waal and Luttrell, 1986).

The number of fights among kin ceases to be exceptional if time spent together is taken into account. The high aggression rate is therefore partly a product of proximity. Yet the same is true of other kinds of interaction, so that a connection with proximity cannot be used to devalue high aggression rates among kin unless one is prepared to do the same for grooming and other affiliative behavior (Bernstein, Judge, and Ruehlmann, 1993).

There is no evidence that aggression among kin is relatively mild: injuries do occasionally result. How, then, does the matrilineal organization survive all this infighting? The key is that fights among kin and other close associates are relatively often reconciled (reviewed by Kappeler and van Schaik, 1992). A parallel report in human children finds friends preserving proximity despite high conflict rates (Hartup et al., 1988).

21. Obviously, juvenile chimpanzees do not vomit as a conscious threat; they simply get extremely frustrated and upset. My "investment" language is

inspired by Trivers' (1974) provocative analysis of parent-offspring conflict as a genetic tug-of-war.

In this battle of wills mothers have weapons, too. Jane Goodall offered a striking example at a symposium entitled "The Great Apes Revisited" (Mexico, 1994). In Gombe National Park, chimpanzee Fifi had shown regular five-year interbirth intervals, almost like clockwork. Her last birth, however, occurred when her previous child, Faustino, was only three and a half years old. The abrupt drop in maternal attention caused Faustino to have incredible tantrums. More than a year too young for weaning, he must have found life definitely "unjust." Fifi did not completely wean him, though; she continued to allow him to sleep in her nest, even to nurse along with his new sibling.

One of Fifi's original answers to Faustino's tantrums was to climb high in a tree and throw him literally to the ground, at the last instant holding onto an ankle. The young male would hang upside down for fifteen seconds or more, screaming wildly, before his mother retrieved him. Fifi was seen to employ this scare tactic twice in a row, after which Faustino stopped having tantrums that day.

22. Charlesworth, 1991, p. 355.
23. Hearne, 1986, pp. 43–44.
24. This discussion is based on the work of Smuts and Watanabe (1990) and of Colmenares (1991). The study of baboon greetings has a much longer history, however, going back to Kummer's (1968) observations of so-called notification in hamadryas baboons. It seems a particularly promising area for research into nonverbal negotiation. Colmenares (1991, p. 59) summarizes the ingredients of negotiation as follows: (a) conflict of interest, (b) attempts to reach nonaggressive solutions, (c) assessing and influencing the other's intentions, and (d) accommodating one's initial goals to those of the other. For further thoughts on animal negotiation see Hinde (1985), Dunbar (1988, pp. 238–248), Noë (1992), and Chadwick-Jones (1992).
25. Smuts and Watanabe, 1990, p. 169.
26. Goodwin, quoted in Haglund, 1992, p. 140.
27. Quoted in *Washington Post,* February 28, 1992.
28. Ardrey, 1970, p. 62.
29. The reported homicide rates for the Netherlands and the United States come from the *World Health Statistics Annual* of 1987 and 1988. Homicide statistics per city were reported in the *New York Times* of August 5, 1990.
30. Thompson, 1976, p. 226.
31. The general public loves to hear about the biological roots of vice and depravity. Ever since de Mandeville (1714), authors have been more than willing to cater to this desire (Montagu, 1968). We first met this genre of literature in Chapter 1, in relation to the evolutionary process, but the

aggression literature has, if possible, produced even more bad news about our species.

The standard trick is to present mean and selfish acts as proof of our true character, and to either overlook human kindness and sympathy, or demonstrate a hidden agenda behind it. Thus, in the same paradoxical way that the sexlessness of a dream can be taken as evidence for sexual repression, the manifest civility of most people can be explained as a facade hiding a brutish nature (Thompson, 1976).

32. Alexander and Roth, 1971, p. 82.

33. Perhaps the closest primate parallel to Calhoun's (1962) expanding rat population was a group of Japanese macaques in a corral of 0.8 hectare that grew from 107 to 192 individuals over a five-year period. Despite this increase of 79 percent in population density, the mean individual rate of aggressive acts per hour did not significantly vary over the years (Eaton, Modahl, and Johnson, 1980). For reviews of crowding research on nonhuman primates see Erwin (1979) and de Waal (1989c).

34. The table below summarizes the effect of a twentyfold area reduction on the Arnhem chimpanzee colony, based on data from Nieuwenhuijsen and de Waal (1982). Weighted individual rates of behavior during two summer periods (outdoors) are compared with those of three winter periods (indoors), resulting in factorial increases. Note that all measures except play were higher during the winters.

Measure	Outdoors	Indoors
Aggression	1	1.7
Proportion of severe aggression	1	1.1
Submissive greetings	1	2.4
Social grooming	1	2.0
Social Play	1.4	1

35. Given that macaques live in permanent groups in the wild, whereas chimpanzees form fission-fusion societies, it is entirely possible that the observed similarity in conciliatory tendency between captive and wild macaques does not generalize to captive and wild chimpanzees: distancing between adversaries *is* a realistic option for wild chimpanzees.

Even though chimpanzees on the Arnhem island have a great deal of space, they cannot get out of each other's way for more than a couple of hours. The observed similarity in conciliatory tendency during indoor and outdoor periods may be due, therefore, to the chimpanzees' having learned that it is better to reconcile right away than to run into an ill-willed opponent later on.

Some of the most spectacular reconciliations in Arnhem took place in the late afternoon, when the chimpanzees were being called into the building for their evening meal. If rival males had avoided each other following a confrontation, they might find themselves in the awkward and dangerous situation of having to spend the night unreconciled together in the same room. Under these circumstances, males were often seen circling each other outside, near the entrance door, testing preparedness to make up. These tense scenes usually ended with an outburst of screaming and embracing between the two—a highly emotional moment followed by grooming. Sometimes they groomed each other so long that the keeper proceeded to feed the rest of the colony while we waited until the two males had calmed down sufficiently to come in together. These reconciliations, clearly forced on the males by the anticipation of proximity, are oriented to the future. Accordingly, I have called them "truces" (de Waal, 1982, pp. 113–114).

36. For chimpanzees there is no such thing as "the" natural habitat. Environments in which these apes survive today range from relatively open, wooded savannas to dense rain forests. Moreover, most known chimpanzee populations have been affected by human activity, such as hunting or food provisioning. On the basis of her reading of the literature, Power (1991) has argued that provisioning at some field sites (such as Gombe's banana camp) turned the chimpanzees more violent and less egalitarian, and thus changed the "tone" of relationships both within and between communities. Power's analysis—which blends a serious reexamination of available data with nostalgia for the 1960s image of apes as noble savages—raises questions that will no doubt be settled by ongoing research on unprovisioned wild chimpanzees.

   The issue of provisioning is also relevant to this chapter's comparison of macaques under different conditions, because all populations (except the one studied in the field by Aureli, 1992) were provisioned. It is not well understood how the behavior of provisioned free-ranging monkeys compares to that of wild monkeys subsisting on dispersed natural foods. Asquith (1989) has reviewed the pros and cons of provisioning in the field.

37. Strier, 1992, p. 70.

38. In light of the current advances in genetic technology, a statement such as "We have to work within our biological endowment" is rapidly becoming outdated.

39. If monkeys sometimes reconcile "for" their matriline, the same mechanism might operate between groups. Intergroup relations were documented in a free-ranging population of rhesus monkeys at Morgan Island, South Carolina. The vast majority of intergroup encounters were hostile, but on eleven separate occasions adult females belonging to *different* groups were seen to assemble for grooming. Remarkably, these contacts involved alpha females of the respective groups, particularly of groups between which much fighting occurred. In a parallel with international diplomacy, several of these high-level contacts took place

shortly after an intergroup fight, and may thus have served to reestablish peaceful relations (Judge and de Waal, 1994).

40. See de Waal (1989a, pp. 107–110) and Cheney and Seyfarth (1990, pp. 72–86) for further discussion of nonegocentric social knowledge.

41. Bodily harm to self is no doubt the first and foremost constraint on competition. According to Maynard Smith and Price (1973), the advantages of winning a fight are weighed against the cost of injury in case the adversary fights back. Van Rhijn and Vodegel (1980) refined this model by including the role of individual recognition.

These early evolutionary models, however, considered physical damage only. Elsewhere I have discussed constraints on competition from the perspective of *social* damage. The social environment may be regarded as a set of resources (for example, cooperative relationships) the effective exploitation of which requires that energy be put into their preservation (de Waal, 1989b). Thus, Sigg and Falett (1985) speculated that the main function of tolerance in relation to food is that it permits high-ranking and low-ranking baboons to forage side by side, which in turn helps all of them to close ranks and defend themselves in times of danger.

6. Conclusion

1. Dewey, 1993 (1898), pp. 109–110.

2. The father of Japanese primatology, Kinji Imanishi, defined culture as "socially transmitted adjustable behavior" (Nishida, 1987, p. 462). Early data and concepts in the study of animal culture may be found in Kummer (1971), Menzel (1973), and Bonner (1980). Cultural transmission is usually contrasted with the acquisition of behavior through individual learning and/or genetic transmission: cultural transmission implies *learning from* others, and in its most effective form (perhaps limited to our species) *teaching by* others. Two recent books reviewing primate cultural phenomena are McGrew (1992) and Wrangham et al. (1994).

3. Proudhon, quoted in Hardin, 1982, p. 184.

4. For a balanced account of the controversy over the use of monkeys and apes in biomedical research, see Blum (1994). This science journalist's book ends with a description of the Yerkes Field Station as the sort of model facility that animal advocates should be able to accept. As a behavioral scientist working at this facility and a few others devoted to research, education, and conservation, I often feel caught in the middle of the controversy. I fully recognize the need to treat animals with respect and compassion, and I am a strong proponent of social housing for nonhuman primates (see, for instance, de Waal, 1992c). At the same time, I greatly appreciate the benefits derived from biomedical research, benefits that even the staunchest critics of such research are not prepared to decline.

The strategy of both sides in this acrimonious debate has been to depict the other as inhumane or immoral. Ironically, a debate such as this one is precisely what defines us as moral beings. As with all true dilemmas, the majority of people feel torn between two undesirables: inflicting pain upon animals and forgoing the medical advances produced thereby.

BIBLIOGRAPHY

Adang, O. M. J. 1984. Teasing in young chimpanzees. *Behaviour* 88:98–122.
——— 1986. Teasing, harassment, provocation: The development of quasi-aggressive behaviour in chimpanzees. Ph.D. diss., University of Utrecht.
Alexander, B., and E. Roth. 1971. The effects of acute crowding on aggressive behavior of Japanese monkeys. *Behaviour* 39:73–89.
Alexander, R. D. 1987. *The Biology of Moral Systems.* New York: Aldine.
——— 1993. Biological considerations in the analysis of morality. In M. H. Nitecki and D. V. Nitecki, eds., *Evolutionary Ethics,* pp. 163–196. Albany: State University of New York Press.
Altmann, J. 1980. *Baboon Mothers and Infants.* Cambridge, Mass.: Harvard University Press.
Altmann, S. A. 1981. Dominance relationships: The Cheshire cat's grin? *Behavioral and Brain Sciences* 4:430–431.
Ardrey, R. A. 1967. *The Territorial Imperative.* London: Collins.
——— 1970. The violent way. *Life,* Sept. 11, pp. 56–68.
Arnhart, L. 1993. How animals move from "is" to "ought": Darwin, Aristotle, and the ethics of desire. Paper presented at the annual meeting of the American Political Science Association, Washington, D.C.
Asquith, P. 1984. The inevitability and utility of anthropomorphism in description of primate behaviour. In R. Harré and V. Reynolds, eds., *The Meaning of Primate Signals,* pp. 138–176. Cambridge: Cambridge University Press.

———— 1989. Provisioning and the study of free-ranging primates: History, effects, and prospects. *Yearbook of Physical Anthropology* 32:129–158.

Atwood, M. E. 1989. *Cat's Eye.* New York: Doubleday.

Aureli, F. 1992. Post-conflict behaviour among wild long-tailed macaques *(Macaca fascicularis). Behavioral Ecology and Sociobiology* 31:329–337.

Aureli, F., and C. P. van Schaik. 1991a. Post-conflict behaviour in long-tailed macaques *(Macaca fascicularis):* I. The social events. *Ethology* 89:89–100.

———— 1991b. Post-conflict behaviour in long-tailed macaques *(Macaca fascicularis):* II. Coping with the uncertainty. *Ethology* 89:101–114.

Aureli, F., et al. 1992. Kin-oriented redirection among Japanese macaques: An expression of a revenge system? *Animal Behaviour* 44:283–291.

———— 1993. Reconciliation, consolation, and redirection in Japanese macaques *(Macaca fuscata). Behaviour* 124:1–21.

———— 1994. Post-conflict social interactions among barbary macaques *(Macaca sylvana). International Journal of Primatology* 15:471–485.

Axelrod, R., and W. D. Hamilton. 1981. The evolution of cooperation. *Science* 211:1390–96.

Badcock, C. R. 1986. *The Problem of Altruism: Freudian-Darwinian Solutions.* Oxford: Blackwell.

Badrian, N., and R. Malenky. 1984. Feeding ecology of *Pan paniscus* in the Lomako Forest, Zaire. In R. L. Susman, ed., *The Pygmy Chimpanzee: Evolutionary Biology and Behavior,* pp. 275–299. New York: Plenum.

Baker, K., and B. B. Smuts. 1994. Social relationships of female chimpanzees: Diversity between captive social groups. In R. W. Wrangham et al., eds., *Chimpanzee Cultures,* pp. 227–242. Cambridge, Mass.: Harvard University Press.

Barlow, C. 1991. *From Gaia to Selfish Genes.* Cambridge, Mass.: MIT Press.

Batson, C. D. 1990. How social an animal: The human capacity for caring. *American Psychologist* 45:336–346.

Bauers, K. A. 1993. A functional analysis of staccato grunt vocalizations in the stumptailed macaque *(Macaca arctoides). Ethology* 94:147–161.

Bercovitch, F. B. 1988. Coalitions, cooperation, and reproductive tactics among adult male baboons. *Animal Behaviour* 36:1198–1209.

Berkson, G. 1973. Social responses to abnormal infant monkeys. *American Journal of Physical Anthropology* 38:583–586.

Bernstein, I. S. 1969. Spontaneous reorganization of a pigtail monkey group. *Proceedings of the 2nd International Congress of Primatology, 1968,* vol. 1, pp. 48–51. Basel: Karger.

———— 1981. Dominance: The baby and the bathwater. *Behavioral and Brain Sciences* 4:419–458.

Bernstein, I. S., and C. Ehardt. 1986. The influence of kinship and socialization on aggressive behavior in rhesus monkeys *(Macaca mulatta)*. *Animal Behaviour* 34:739–747.

Bernstein, I. S., P. G. Judge, and T. E. Ruehlmann. 1993. Kinship, association, and social relationships in rhesus monkeys. *American Journal of Primatology* 31:41–54.

Betzig, L. L. 1986. *Despotism and Differential Reproduction: A Darwinian View of History.* New York: Aldine.

Birch, L. L., and J. Billman. 1986. Preschool children's food sharing with friends and acquaintances. *Child Development* 57:387–395.

Bischof-Köhler, D. 1988. Über den Zusammenhang von Empathie und der Fähigkeit sich im Spiegel zu erkennen. *Schweizerische Zeitschrift für Psychologie* 47:147–159.

Blum, D. 1994. *The Monkey Wars.* New York: Oxford University Press.

Blurton Jones, N. G. 1987. Tolerated theft, suggestions about the ecology and evolution of sharing, hoarding and scrounging. *Social Science Information* 26(1):31–54.

Boehm, C. 1992. Segmentary warfare and the management of conflict: Comparison of East African chimpanzees and patrilineal-patrilocal humans. In A. H. Harcourt and F. B. M. de Waal, eds., *Coalitions and Alliances in Humans and Other Animals,* pp. 137–173. Oxford: Oxford University Press.

—— 1993. Egalitarian behavior and reverse dominance hierarchy. *Current Anthropology* 34:227–254.

Boesch, C. 1991a. Teaching among wild chimpanzees. *Animal Behaviour* 41:530–532.

—— 1991b. The effects of leopard predation on grouping patterns in forest chimpanzees. *Behaviour* 117:220–242.

—— 1992. New elements of a theory of mind in wild chimpanzees. *Behavioral and Brain Sciences* 15:149–150.

—— 1993. Towards a new image of culture in wild chimpanzees? *Behavioral and Brain Sciences* 16 :514–515.

—— 1994a. Chimpanzee—red colobus monkeys: A predator-prey system. *Animal Behaviour* 47:1135–48.

—— 1994b. Cooperative hunting in wild chimpanzees. *Animal Behaviour* 48:653–667.

Boesch, C., and H. Boesch. 1989. Hunting behavior of wild chimpanzees in the Taï National Park. *American Journal of Physical Anthropology* 78:547–573.

Boinski, S. 1988. Use of a club by a wild white-faced capuchin *(Cebus capucinus)* to attack a venomous snake *(Bothrops asper)*. *American Journal of Primatology* 14:177–179.

—— 1992. Monkeys with inflated sex appeal. *Natural History* 101 (7):42–48.

Bonner, J. T. 1980. *The Evolution of Culture in Animals.* Princeton: Princeton University Press.

Bowlby, J. 1983 (1973). *Attachment and Loss,* vol. 2, *Separation Anxiety and Anger.* Harmondsworth: Penguin.

van Bree, P. 1963. On a specimen of *Pan paniscus,* Schwarz 1929, which lived in the Amsterdam Zoo from 1911 till 1916. *Zoologische Garten* 27:292–295.

Breggin, P. R. 1992. We could all learn a thing or two from the chimpanzees. *New York Times,* Mar. 15.

Breuggeman, J. A. 1973. Parental care in a group of free-ranging rhesus monkeys *(Macaca mulatta). Folia primatologica* 20:178–210.

Brigham J. C., and R. S. Malpass. 1985. The role of experience and contact in the recognition of faces of own- and other-race persons. *Journal of Social Issues* 41:139–155.

Brothers, L. 1989. A biological perspective on empathy. *American Journal of Psychiatry* 146:10–19.

Broughton, J. M. 1983. Woman's rationality and men's virtues: A critique of gender dualism in Gilligan's theory of moral development. *Social Research* 50:597–642.

Burghardt, G. M. 1985. Animal awareness: Current perceptions and historical perspective. *American Psychologist* 40:905–919.

——— 1991. Cognitive ethology and critical anthropomorphism: A snake with two heads and hog-nosed snakes that play dead. In C. A. Risteau, ed., *Cognitive Ethology,* pp. 55–90. Hillsdale, N.J.: Erlbaum.

——— 1992. Human-bear bonding in research on black bear behavior. In H. Davis and D. Balfour, eds., *The Inevitable Bond: Examining Scientist-Animal Interactions,* pp. 365–382. Cambridge: Cambridge University Press.

Buttersworth, G. E., et al. 1991. *Perspectives on the Child's Theory of Mind.* Oxford: Oxford University Press.

Bygott, J. D. 1979. Agonistic behavior, dominance, and social structure in wild chimpanzees of the Gombe National Park. In D. A. Hamburg and E. R. McCown, eds., *The Great Apes,* pp. 405–428. Menlo Park, Calif.: Benjamin Cummings.

Byrne, R. W. 1995. *The Thinking Ape.* Oxford: Oxford University Press.

Byrne, R. W., and A. Whiten. 1988. *Machiavellian Intelligence: Social Expertise and the Evolution of Intellect in Monkeys, Apes, and Humans.* Oxford: Oxford University Press.

——— 1990. Tactical deception in primates: The 1990 database. *Primate Report* 27:1–101.

Caine, N., and M. Reite. 1983. Infant abuse in captive pig-tailed macaques: Relevance to human child abuse. In M. Reite and N. Caine, eds., *Child Abuse: The Nonhuman Primate Data,* pp. 19–27. New York: Liss.

Caldwell, M. C., and D. K. Caldwell. 1966. Epimeletic (care-giving) behavior in Cetacea. In K. S. Norris, ed., *Whales, Dolphins, and Porpoises*, pp. 755–789. Berkeley: University of California Press.

Calhoun, J. B. 1962. Population density and social pathology. *Scientific American* 206:139–148.

Campbell, D. T. 1975. On the conflicts between biological and social evolution and between psychology and moral tradition. *American Psychologist* 117:1103–26.

Canetti, E. 1963 (1960). *Crowds and Power.* New York: Viking Press.

Caporael, L. R., et al. 1989. Selfishness examined: Cooperation in the absence of egoistic incentives. *Behavioral and Brain Sciences* 12:683–739.

Carey, S. 1985. *Conceptual Change in Childhood.* Cambridge, Mass.: MIT Press.

Carpenter, E. 1975. The tribal terror of self-awareness. In P. Hockings, ed., *Principles of Visual Anthropology*, pp. 451–461. The Hague: Mouton.

Cartmill, M. 1993. *A View to a Death in the Morning: Hunting and Nature through History.* Cambridge, Mass.: Harvard University Press.

Cavalieri, P., and P. Singer. 1993. *The Great Ape Project: Equality beyond Humanity.* London: Fourth Estate.

Cenami Spada E., et al. Forthcoming. The self as reference point: Can animals do without it? In P. Rochat, ed., *The Self in Infancy: Theory and Research.* Amsterdam: Elsevier.

Chadwick-Jones, J. K. 1992. Baboon social communication: The social contingency model and social exchange. In F. D. Burton, ed., *Social Processes and Mental Abilities in Non-Human Primates*, pp. 91–108. Lewiston, N.Y.: Mellen.

Chagnon, N. A. 1988. Life histories, blood revenge, and warfare in a tribal population. *Science* 239:985–992.

Chance, M., G. Emory, and R. Payne. 1977. Status referents in long-tailed macaques *(Macaca fascicularis):* Precursors and effects of a female rebellion. *Primates* 18:611–632.

Charlesworth, W. R. 1991. The development of the sense of justice. *American Behavioral Scientist* 34:350–370.

Cheney, D. L., and R. M. Seyfarth. 1986. The recognition of social alliances by vervet monkeys. *Animal Behaviour* 34:1722–31.

———— 1989. Reconciliation and redirected aggression in vervet monkeys, *Cercopithecus aethiops. Behaviour* 110:258–275.

———— 1990. *How Monkeys See the World: Inside the Mind of Another Species.* Chicago: University of Chicago Press.

———— 1991. Reading minds or reading behaviour? Tests of a theory of mind in monkeys. In A. Whiten, ed., *Natural Theories of Mind; Evolution, Development and Simulation of Everyday Mindreading*, pp. 175–194. Oxford: Blackwell.

Ci, J. 1991. Conscience, sympathy and the foundation of morality. *American Philosophical Quarterly* 28:49–59.

Clutton-Brock, T. H., and G. A. Parker. 1995. Punishment in animal societies. *Nature* 373:209–216.

Coe, C. L., and L. A. Rosenblum. 1984. Male dominance in the bonnet macaque: A malleable relationship. In P. R. Barchas and S. P. Mendoza, eds., *Social Cohesion: Essays toward a Sociophysiological Perspective*, pp. 31–63. Westport, Conn.: Greenwood.

Colby, A., and W. Damon. 1983. Listening to a different voice: A review of Gilligan's *In a Different Voice. Merrill-Palmer Quarterly* 29:473–481.

Colmenares, F. 1991. Greeting behaviour between male baboons: Oestrus females, rivalry, and negotiation. *Animal Behaviour* 41:49–60.

Connor, R. C., and K. S. Norris. 1982. Are dolphins reciprocal altruists? *American Naturalist* 119:358–372.

Cords, M. 1988. Resolution of aggressive conflicts by immature male long-tailed macaques. *Animal Behaviour* 36:1124–35.

Cords, M., and S. Thurnheer. 1993. Reconciling with valuable partners by long-tailed macaques. *Ethology* 93:315–325.

Coser, L. A. 1966. Some social functions of violence. *Annals* 364:8–18.

Cronin, H. 1991. *The Ant and the Peacock*. Cambridge: Cambridge University Press.

Crook, J. H. 1989. Introduction: Socioecological paradigms, evolution and history: Perspectives for the 1990s. In V. Standen and R. A. Foley, eds., *Comparative Socioecology: The Behavioural Ecology of Humans and Other Mammals*, pp. 1–36. London: Blackwell.

Damasio, H., et al. 1994. The return of Phineas Gage: Clues about the brain from the skull of a famous patient. *Science* 264:1102–05.

D'Amato, M. R., and N. Eisenstein. 1972. Laboratory breeding and rearing of *Cebus apella. Laboratory Primate Newsletter* 11(3):4–9.

Darley, J. M., and C. D. Batson. 1973. From Jerusalem to Jericho: A study of situational and dispositional variables in helping behavior. *Journal of Personality and Social Psychology* 27:100–108.

Dart, R. A. 1959. *Adventures with the Missing Link*. New York: Harper.

Darwin, C. 1965 (1872). *The Expression of the Emotions in Man and Animals*. Chicago: University of Chicago Press.

——— 1981 (1871). *The Descent of Man, and Selection in Relation to Sex*. Princeton: Princeton University Press.

Dasser, V. 1988. A social concept in Java-monkeys. *Animal Behaviour* 36:225–230.

Davis, H. 1989. Theoretical note on the moral development of rats *(Rattus norvegicus). Journal of Comparative Psychology* 103:88–90.

Dawe, H. C. 1934. An analysis of two hundred quarrels of preschool children. *Child Development* 5:139–157.

Dawkins, R. 1976. *The Selfish Gene.* Oxford: Oxford University Press.

——— 1981. In defense of selfish genes. *Philosophy* 56:556–573.

Demaria, C., and B. Thierry. 1990. Formal biting in stump-tailed macaques *(Macaca arctoides). American Journal of Primatology* 20:133–140.

Deputte, B. L. 1988. Perception de la mort et de la séparation chez les primates. *Nouvelle Revue d'Ethnopsychiatrie* 10:113–150.

Dettwyler, K. A. 1991. Can paleopathology provide evidence for "compassion"? *American Journal of Physical Anthropology* 84:375–384.

Dewey, J. 1993 (1898). Evolution and ethics. Reprinted in M. H. Nitecki and D. V. Nitecki, eds., *Evolutionary Ethics,* pp. 95–110. Albany: State University of New York Press.

Dickey, L. 1986. Historicizing the "Adam Smith Problem": Conceptual, historiographical, and textual issues. *Journal of Modern History* 58:579–609.

Dittus, W. P. J. 1979. The evolution of behaviors regulating density and age-specific sex ratios in a primate population. *Behaviour* 69:265–302.

Dittus, W. P. J., and S. M. Ratnayeke. 1989. Individual and social behavioral responses to injury in wild toque macaques *(Macaca sinica). International Journal of Primatology* 10:215–234.

Dugatkin, L. A. 1992. Sexual selection and imitation: Females copy the mate choice of others. *American Naturalist* 139:1384–89.

Dunbar, R. I. M. 1988. *Primate Social Systems.* London: Croom Helm.

Dutton, D., and S. L. Painter. 1981. Traumatic bonding: The development of emotional attachments in battered women and other relationships of intermittent abuse. *Victimology* 6:139–155.

Eaton, G., K. B. Modahl, and D. F. Johnson. 1980. Aggressive behavior in a confined troop of Japanese macaques: Effects of density, season, and gender. *Aggressive Behavior* 7:145–164.

Edwards, C. P. 1993. Behavioral sex differences in children of diverse cultures: The case of nurturance to infants. In M. E. Pereira and L. A. Fairbanks, eds., *Juvenile Primates: Life History, Development, and Behavior,* pp. 327–338. New York: Oxford University Press.

Ehardt, C. L., and I. S. Bernstein. 1992. Conflict intervention behaviour by adult male macaques: Structural and functional aspects. In A. H. Harcourt and F. B. M. de Waal, eds., *Coalitions and Alliances in Humans and Other Animals,* pp. 83–111. Oxford: Oxford University Press.

Eibl-Eibesfeldt, I. 1990. Dominance, submission, and love: Sexual pathologies from the perspective of ethology. In J. R. Feierman, ed., *Pedophilia: Biosocial Dimensions,* pp. 150–175. New York: Springer.

Erdal, D., and A. Whiten. 1994. On human egalitarianism: An evolutionary product of Machiavellian status escalation? *Current Anthropology,* 35:175–183.

——— In press. Egalitarianism and Machiavellian intelligence in human evolution. *Cambridge Archeological Journal.*

Erwin, J. 1979. Aggression in captive macaques: Interaction of social and spatial factors. In J. Erwin, T. L. Maple, and G. Mitchell, eds., *Captivity and Behavior,* pp. 139–171. New York: Van Nostrand.

Fedigan, L. M., and L. Fedigan. 1977. The social development of a handicapped infant in a free-living troop of Japanese monkeys. In S. Chevalier-Skolnikoff and F. E. Poirier, eds., *Primate Bio-Social Development: Biological, Social, and Ecological Determinants,* pp. 205–222. New York: Garland.

Feistner, A. T. C., and W. C. McGrew. 1989. Food-sharing in primates: A critical review. In P. K. Seth and S. Seth, eds., *Perspectives in Primate Biology,* vol. 3, pp. 21–36. New Delhi: Today and Tomorrow's Printers and Publishers.

Fiorito, G., and P. Scotto. 1992. Observational learning in *Octopus vulgaris. Science* 256:545–547.

Fischler, C. 1990. *L'Homnivore: Le goût, la cuisine et le corps.* Paris: Odile Jacob.

Fisher, A. 1991. A new synthesis comes of age. *Mosaic* 22:3–17.

Fossey, D. 1983. *Gorillas in the Mist.* Boston: Houghton Mifflin.

Frank, R. H. 1988. *Passions within Reason: The Strategic Role of the Emotions.* New York: Norton.

Freedman, D. G. 1958. Constitutional and environmental interactions in rearing of four breeds of dogs. *Science* 127:585–586.

Freud, S. 1961 (1930). *Civilization and its Discontents.* New York: Norton.

Galef, B. G. 1988. Imitation in animals: History, definitions, and interpretation of data from the psychological laboratory. In T. Zentall and B. G. Galef, eds., *Psychological and Biological Perspectives,* pp. 1–28. Hillsdale N.J.: Erlbaum.

Gallup, G. 1982. Self-awareness and the emergence of mind in primates. *American Journal of Primatology* 2:237–248.

Garner, R. L. 1896. *Gorillas and Chimpanzees.* London: Osgood.

Gibber, J. R. 1981. Infant-directed behaviors in male and female rhesus monkeys. Ph.D. diss., University of Wisconsin–Madison.

Gibson, E. J. 1994. Has psychology a future? *Psychological Science* 5:69–76.

Gibson, J. J. 1979. *The Ecological Approach to Visual Perception.* Boston: Houghton Mifflin.

Gilligan, C. 1982. *In a Different Voice: Psychological Theory and the Women's Movement.* Cambridge, Mass.: Harvard University Press.

Ginsberg, H., and S. Miller. 1981. Altruism in children: A naturalistic study of reciprocation and an examination of the relationship between social dominance and aid-giving behavior. *Ethology and Sociobiology* 2:75–83.

Goldfoot, D. A., and D. A. Neff. 1985. On measuring sex differences in social contexts. In N. Adler, D. Pfaff, and R. W. Goy, eds., *Handbook of Behavioral Neurobiology,* vol. 7, pp. 767–783. New York: Plenum.

Goodall, J. 1971. *In the Shadow of Man.* Boston: Houghton Mifflin.

—— 1986. *The Chimpanzees of Gombe: Patterns of Behavior.* Cambridge, Mass.: Belknap Press, Harvard University Press.

—— 1990. *Through a Window.* Boston: Houghton Mifflin.

—— 1992. Unusual violence in the overthrow of an alpha male chimpanzee at Gombe. In T. Nishida, et. al., eds., *topics in Primatology,* vol. 1, *Human Origins,* pp. 131–142. Tokyo: University of Tokyo Press.

Gottman, J. 1994. *Why Marriages Succeed or Fail.* New York: Simon and Schuster.

Gould, S. J. 1980. So cleverly kind an animal. In *Ever since Darwin,* pp. 260–267. Harmondsworth: Penguin.

—— 1988. Kropotkin was no crackpot. *Natural History* 97(7):12–21.

Grammer, K. 1992. Intervention in conflicts among children: Contexts and consequences. In: A. H. Harcourt and F. B. M. de Waal, eds., *Coalitions and Alliances in Humans and other Animals,* pp. 258—283. Oxford: Oxford University Press.

Guisso, R. W. L., and C. Pagani. 1989. *The First Emperor of China.* New York: Birch Lane.

Hadwin, J., and J. Perner. 1991. Pleased and surprised: Children's cognitive theory of emotion. *British Journal of Developmental Psychology* 9:215–234.

Haglund, K. 1992. Violence research is due for attention. *Journal of NIH Research* 4:38–42.

Hall, K. R. L. 1964. Aggression in monkey and ape societies. In J. Carthy and F. Ebling, eds., *The Natural History of Aggression,* pp. 51–64. London: Academic Press.

Hamilton, W. D. 1971. Selection of selfish and altruistic behavior. In J. F. Eisenberg and W. S. Dillon, eds., *Man and Beast: Comparative Social Behavior,* pp. 59–91. Washington, D.C.: Smithsonian Institution Press.

Harcourt, A. H., and F. B. M. de Waal. 1992. *Coalitions and Alliances in Humans and Other Animals.* Oxford: Oxford University Press.

Hardin, G. 1982. Discriminating altruisms. *Zygon* 17:163–186.

Hartup, W., et al. 1988. Conflict and the friendship relations of young children. *Child Development* 59:1590–1600.

Hatfield, E., J. T. Cacioppo, and R. L. Rapson. 1993. Emotional contagion. *Current Directions in Psychological Science* 2:96–99.

Hawkes, K. 1990. Showing off: Tests of an hypothesis about men's foraging goals. *Ethology and Sociobiology* 12:29–54.

Hearne, V. 1986. Reflections: Questions about language. I—Horses. *New Yorker* 62 (Aug.):33–57.

Hediger, H. 1955. *Studies in the Psychology and Behaviour of Animals in Zoos and Circuses.* London: Butterworth.

Heine, B. 1985. The mountain people: Some notes on the Ik of north-eastern Uganda. *Africa* 55:3–16.

Heinrich, B. 1989. *Ravens in Winter.* New York: Summit.

Heyes, C. M. 1993. Anecdotes, training, trapping and triangulating: Do animals attribute mental states? *Animal Behaviour* 46:177–188.

Hinde, R. A. 1970. Aggression. In J. Pringle, ed., *Biology and the Human Sciences,* pp. 1–23. Oxford: Clarendon.

—— 1985. Expression and negotiation. In G. Zivin, ed., *The Development of Expressive Behavior: Biology-Environment Interactions,* pp. 103–116. Orlando, Fla.: Academic Press.

Hobbes, T. 1991 (1651). *Leviathan.* Cambridge: Cambridge University Press.

Hoffman, M. L. 1978. Sex differences in empathy and related behaviors. *Psychological Bulletin* 84:712–722.

—— 1981. Perspectives on the difference between understanding people and understanding things: The role of affect. In J. H. Flavell and L. Ross, eds., *Social Cognitive Development,* pp. 67–81. Cambridge: Cambridge University Press.

Hohmann, G., and B. Fruth. 1993. Field observations on meat sharing among bonobos *(Pan paniscus). Folia primatologica* 60:225–229.

—— Forthcoming. Food sharing and status in provisioned bonobos *(Pan paniscus):* Preliminary results. In P. Wiessner and W. Schiefenhövel, eds., *Food and the Status Quest.* Oxford: Berghahn.

Holldöbbler, B., and E. O. Wilson. 1990. *The Ants.* Cambridge, Mass.: Harvard University Press.

Hopkins, W. D., and R. D. Morris. 1993. Handedness in great apes: A review of findings. *International Journal of Primatology* 14:1–25.

Howell, M., J. Kinsey, and M. A. Novak. 1994. Mark-directed behavior in a rhesus monkey after controlled reinforced exposure to mirrors. *American Journal of Primatology* 33:216.

Hume, D. 1978 (1739). *A Treatise of Human Nature.* Oxford: Oxford University Press.

Huxley, T. H. 1888. Struggle for existence and its bearing upon man. *Nineteenth Century,* Feb., 1888.

—— 1989 (1894). *Evolution and Ethics.* Princeton: Princeton University Press.

Ihobe, H. 1990. Interspecific interactions between wild pygmy chimpanzees *(Pan paniscus)* and red colobus *(Colobus badius). Primates* 3:109–112.

—— 1992. Observations on the meat-eating behavior of wild bonobos *(Pan paniscus)* at Wamba, Republic of Zaire. *Primates* 33:247–250.

Isaac, G. 1978. The food-sharing behavior of protohuman hominids. *Scientific American* 238(4):90–108.

Itani, J. 1988. The origin of human equality. In M. R. A. Chance, ed., *Social Fabrics of the Mind,* pp. 137–156. Hove: Erlbaum.

Izawa, K. 1978. Frog-eating behavior of wild black-capped capuchin *(Cebus apella). Primates* 19:633–642.

Jacoby, S. 1983. *Wild Justice: The Evolution of Revenge.* New York: Harper and Row.

Janson, C. H. 1988. Food competition in brown capuchin monkeys *(Cebus apella):* Quantitative effects of group size and tree productivity. *Behaviour* 105:53–76.

Jaynes, J. 1969. The historical origins of "ethology" and "comparative psychology." *Animal Behaviour* 17:601–606.

Jerison, H. J. 1986. The perceptual worlds of dolphins. In R. J. Schusterman, J. A. Thomas, and F. G. Wood, eds., *Dolphin Cognition and Behavior: A Comparative Approach,* pp. 141–166. London: Erlbaum.

Johnson, D. B. 1982. Altruistic behavior and the development of self in infants. *Merrill-Palmer Quarterly* 28:379–388.

Johnston, K. D. 1988. Adolescents' solutions to dilemmas in fables: Two moral orientations—two problem-solving strategies. In C. Gilligan, J. V. Ward, and M. Taylor, eds., *Mapping the Moral Domain: A Contribution of Women's Thinking to Psychological Theory and Education,* pp. 49–71. Cambridge, Mass.: Harvard University Press.

Joubert, D. 1991. Elephant wake. *National Geographic* 179:39–42.

Judge, P. G. 1991. Dyadic and triadic reconciliation in pigtail macaques *(Macaca nemestrina). American Journal of Primatology* 23:225–237.

Judge, P. G., and F. B. M. de Waal. 1994. Intergroup grooming relations between alpha females in a population of free-ranging rhesus macaques. *Folia primatologica* 63:63–70.

Kant, I. 1947 (1785). Groundwork of the metaphysic of morals. In H. J. Paton, trans., *The Moral Law.* London: Hutchinson.

Kaplan, H., and K. Hill. 1985. Hunting ability and reproductive success among Ache foragers: Preliminary results. *Current Anthropology* 26:131–133.

Kappeler, P. M., and C. P. van Schaik. 1992. Methodological and evolutionary aspects of reconciliation among primates. *Ethology* 92:51–69.

Kaufman I. C., and L. A. Rosenblum. 1967. Depression in infant monkeys separated from their mothers. *Science* 155:1030–31.

Kennedy, J. S. 1992. *The New Anthropomorphism.* Cambridge: Cambridge University Press.

Killen, M., and E. Turiel. 1991. Conflict resolution in preschool social interactions. *Early Education and Development* 2:240–255.

Kitcher, P. 1985. *Vaulting Ambition: Sociobiology and the Quest for Human Nature.* Cambridge, Mass.: MIT Press.

Knauft, B. M. 1991. Violence and sociality in human evolution. *Current Anthropology* 32:391–428.

Kohlberg, L. 1984. *Essays on Moral Development,* vol. 2, *The Psychology of Moral Development.* San Fransisco: Harper and Row.

Köhler, W. 1925. *The Mentality of Apes.* New York: Vintage.

Kortlandt, A. 1991. Primates in the looking glass: A historical note. *Animal Behaviour Society Newsletter* 37:11.

Kropotkin, P. 1972 (1902). *Mutual Aid: A Factor of Evolution*. New York: New York University Press.

Kummer, H. 1968. *Social Organization in Hamadryas Baboons*. Chicago: University of Chicago Press.

——— 1971. *Primate Societies*. Chicago: University of Chicago Press.

——— 1978. On the value of social relationships to nonhuman primates: A heuristic scheme. *Social Science Information* 17:687–705.

——— 1982. Social knowledge in free-ranging primates. In D. Griffin, ed., *Animal Mind—Human Mind*, pp. 113–130. Berlin: Springer.

——— 1991. Evolutionary transformations of possessive behavior. In F. W. Rudmin, ed., *To Have Possessions: A Handbook on Ownership and Property*. Special issue of *Journal of Social Behavior and Personality* 6(6):75–83.

Kummer, H., and M. Cords. 1991. Cues of ownership in long-tailed macaques, *Macaca fascicularis*. *Animal Behaviour* 42:529–549.

Kummer, H., and J. Goodall. 1985. Conditions of innovative behaviour in primates. In L. Weiskrantz, ed., *Animal Intelligence*, pp. 203–214. Oxford: Clarendon.

Kummer, H., V. Dasser, and P. Hoyningen-Huene. 1990. Exploring primate social cognition: Some critical remarks. *Behaviour* 112:84–98.

Kummer, H., W. Götz, and W. Angst. 1974. Triadic differentiation: A process protecting pair-bonds in baboons. *Behavior* 49:62–87.

Kurland, J. A. 1977. *Contributions to Primatology*, vol. 12, *Kin Selection in the Japanese Monkey*. Basel: Karger.

Kuroda, S. 1984. Interactions over food among pygmy chimpanzees. In R. Susman, ed., *The Pygmy Chimpanzee*, pp. 301–324. New York: Plenum.

Kyes, R. C. 1992. Protection of a disabled group member in hamadryas baboons. *Laboratory Primate Newsletter* 31:9–10.

Lagerspetz, K. M., K. Björkqvist, and T. Peltonen. 1988. Is indirect aggression typical of females? Gender differences in aggressiveness in 11- to 12-year-old children. *Aggressive Behavior* 14:403–414.

Lee, R. B. 1969. Eating Christmas in the Kalahari. *Natural History* 78(12):14–22, 60–63.

Lee, V. 1913. *The Beautiful: An Introduction to Psychological Aesthetics*. Cambridge: Cambridge University Press.

Lethmate, J. 1977. *Fortschritte der Verhaltensforschung*, vol. 19, *Problemlöseverhalten von Orang-Utans (Pongo pygmaeus)*. Berlin: Paul Parey.

Lethmate, J., and G. Dücker. 1973. Untersuchungen zum Selbsterkennen im Spiegel bei Orang-Utans und einigen anderen Affenarten. *Zeitschrift für Tierpsychogologie* 33:248–269.

Lever, J. 1976. Sex differences in the games children play. *Social Problems* 23:478–487.

Lewin, R. 1984 (1977). Biological limits to morality. In G. Ferry, ed., *The Understanding of Animals*, pp. 226–234. Oxford: Blackwell.

Lieberman, P. 1991. *Uniquely Human: The Evolution of Speech, Thought, and Selfless Behavior.* Cambridge, Mass.: Harvard University Press.

Lippa, R. A. 1990. *Introduction to Social Psychology.* Belmont, Calif.: Wadsworth.

Lopez, B. H. 1978. *Of Wolves and Men.* New York: Scribner's.

Lorenz, K. Z. 1960. *So kam der Mensch auf den Hund.* Vienna: Borotha-Schoeler.

——— 1966 (1963). *On Aggression.* London: Methuen.

——— 1981. *The Foundations of Ethology.* New York: Touchstone.

Lux, K. 1990. *Adam Smith's Mistake.* Boston: Shambhala.

Maccoby, E., and C. Jacklin. 1974. *The Psychology of Sex Differences.* Stanford: Stanford University Press.

de Mandeville, B. 1966 (1714). *The Fable of the Bees: or, Private Vices, Public Benefits,* vol. 1. London: Oxford University Press.

Manning, J. T., R. Heaton, and A. T. Chamberlain. 1994. Left-side cradling: Similarities and differences between apes and humans. *Journal of Human Evolution* 26:77–83.

Marino, L., D. Reiss, and G. G. Gallup. 1994. Mirror self-recognition in bottlenose dolphins: Implications for comparative investigations of highly dissimilar species. In S. T. Parker, R. W. Mitchell, and M. L. Boccia, eds., *Self-Awareness in Animals and Humans: Developmental Perspectives,* pp. 380–391. Cambridge: Cambridge University Press.

Marshall Thomas, E. 1993. *The Hidden Life of Dogs.* Boston: Houghton Mifflin.

Marten, K., and S. Psarakos. 1994. Evidence of self-awareness in the bottlenose dolphin *(Tursiops truncatus).* In S. T. Parker, R. W. Mitchell, and M. L. Boccia, eds., *Self-Awareness in Animals and Humans: Developmental Perspectives,* pp. 361–379. Cambridge: Cambridge University Press.

Maslow, A. H. 1936. The role of dominance in the social and sexual behavior of infra-human primates. I. Observations at Vilas Park Zoo. *Journal of Genetical Psychology* 48:261–277.

——— 1940. Dominance-quality and social behavior in infra-human primates. *Journal of Social Psychology* 11:313–324.

Mason, W. A. 1993. The nature of social conflict: The psycho-ethological perspective. In W. A. Mason and S. P. Mendoza, eds., *Primate Social Conflict,* pp. 13–47. Albany: State University of New York Press.

Masson, J. M., and S. McCarthy. 1995. *When Elephants Weep: The Emotional Lives of Animals.* New York: Delacorte.

Masters, R. 1984. Review of *Chimpanzee Politics*. *Politics and the Life Sciences* 2:208–209.

Matsuzawa, T. 1989. Spontaneous pattern construction in a chimpanzee. In P. Heltne and L. Marquardt, eds., *Understanding Chimpanzees*, pp. 252–265. Cambridge, Mass.: Harvard University Press.

—— 1990. Form perception and visual acuity in a chimpanzee. *Folia primatologica* 55:24–32.

Maynard Smith, J., and G. R. Price. 1973. The logic of animal conflict. *Nature* 246:15–18.

Mayr, E. 1963. *Populations, Species, and Evolution*. Cambridge, Mass.: Belknap Press, Harvard University Press.

McBride, A. F., and D. O. Hebb. 1948. Behavior of the captive bottle-nose dolphin, *Tursiops truncatus*. *Journal of Comparative Physiological Psychology* 41:111–123.

McGrew, W. C. 1975. Patterns of plant food sharing by wild chimpanzees. In *Contemporary Biology 5th International Congress of Primatology, Nagoya 1974*, pp. 304–309. Basel: Karger.

—— 1992. *Chimpanzee Material Culture*. Cambridge: Cambridge University Press.

McGuire, M., M. Raleigh, and C. Johnson. 1983. Social dominance in adult male vervet monkeys: General considerations. *Social Science Information* 22:89–123.

Mednick, M. T. 1989. On the politics of psychological constructs: Stop the bandwagon, I want to get off. *American Psychologist* 44:1118–23.

Menzel, E. W. 1973. *Precultural Primate Behavior*. Basel: Karger.

—— 1974. A group of young chimpanzees in a one-acre field. In A. M. Schrier and F. Stollnitz, eds., *Behavior of Non-human Primates*, vol. 5, pp. 83–153. New York: Academic Press.

—— 1986. How can you tell if an animal is intelligent? In R. J. Schusterman, J. A. Thomas, and F. G. Wood, eds., *Dolphin Cognition and Behavior: A Comparative Approach*, pp.167–181. London: Erlbaum.

—— 1988. Mindless behaviorism, bodiless cognitivism, or primatology? *Behavioral and Brain Sciences* 11:258–259.

Mercer, P. 1972. *Sympathy and Ethics: A Study of the Relationship between Sympathy and Morality with Special Reference to Hume's* Treatise. Oxford: Clarendon.

Midgley, M. 1979. Gene-juggling. *Philosophy* 54:439–458.

—— 1991. *Can't We Make Moral Judgements?* New York: St. Martin's.

—— 1994. *The Ethical Primate: Humans, Freedom and Morality*. London: Routledge.

Miller, B. L. 1990. Aggression and other social behaviors in a captive group of Japanese Monkeys, *Macaca fuscata fuscata*. Master's thesis, University of Wisconsin–Milwaukee.

Milton, K. 1985. Mating patterns of woolly spider monkeys, *Brachyteles arachnoides:* Implications for female choice. *Behavioral Ecology and Sociobiology* 17:53–59.

——— 1992. Civilization and its discontents. *Natural History* 101(3):37–42.

Mitchell, R. W. 1993. Mental models of mirror-self-recognition: Two theories. *New Ideas in Psychology* 11:295–325.

Mitchell, R. W., N. Thompson, and L. Miles. Forthcoming. *Anthropomorphism, Anecdotes, and Animals.* Albany: State University of New York Press.

Montagu, M. F. A. 1968. The new litany of "innate depravity" or original sin revisited. In M. F. A. Montagu, ed., *Man and Aggression.* London: Oxford University Press.

Morgan, E. 1982. *The Aquatic Ape.* New York: Stein and Day.

Moss, C. 1988. *Elephant Memories: Thirteen Years in the Life of an Elephant Family.* New York: Fawcett Columbine.

Nieuwenhuijsen, C., and F. B. M. de Waal. 1982. Effects of spatial crowding on social behavior in a chimpanzee colony. *Zoo Biology* 1:5–28.

Nishida, T. 1968. The social group of wild chimpanzees in the Mahale Mountains. *Primates* 9:167–224.

——— 1979. The social structure of chimpanzees in the Mahale Mountains. In D. A. Hamburg and E. R. McCown, eds., *The Great Apes,* pp. 73–121. Menlo Park, Calif.: Benjamin Cummings.

——— 1983. Alpha status and agonistic alliance in wild chimpanzees. *Primates* 24:318–336.

——— 1987. Local traditions and cultural transmission. In B. B. Smuts et al., eds., *Primate Societies,* pp. 462–474. Chicago: University of Chicago Press.

——— 1989. Social interactions between resident and immigrant female chimpanzees. In P. Heltne and L. Marquardt, eds., *Understanding Chimpanzees,* pp. 68–89. Cambridge, Mass.: Harvard University Press.

——— 1994. Review of recent findings on Mahale chimpanzees. In R. W. Wrangham et al. eds., Chimpanzee Cultures, pp. 373–396. Cambridge, Mass.: Harvard University Press.

Nishida, T., et al. 1992. Meat-sharing as a coalition strategy by an alpha male chimpanzee? In T. Nishida et al., eds., *Topics in Primatology,* vol. 1, *Human Origins,* pp. 159–174. Tokyo: University of Tokyo Press.

Nissen, H. W., and M. P. Crawford. 1936. A preliminary study of food-sharing behavior in young chimpanzees. *Journal of Comparative Psychology* 22:383–419.

Nitecki, M. H., and D. V. Nitecki. 1993. *Evolutionary Ethics.* Albany: State University of New York Press.

Noë, R. 1992. Alliance formation among male baboons: Shopping for profitable partners. In A. H. Harcourt and F. B. M. de Waal, eds., *Coa-*

litions and Alliances in Humans and Other Animals, pp. 285–321. Oxford: Oxford University Press.

Noë, R., F. B. M. de Waal, and J. A. R. A. M. van Hooff. 1980. Types of dominance in a chimpanzee colony. *Folia primatologica* 34:90–110.

Norris, K. S. 1991. *Dolphin Days.* New York: Avon.

Novak, M. A., and H. F. Harlow. 1975. Social recovery of monkeys isolated for the first year of life. I. Rehabilitation and therapy. *Developmental Psychology* 11:453–465.

Nucci, L. P., and E. Turiel. 1978. Social interactions and the development of social concepts in preschool children. *Child Development* 49:400–407.

Parker, S. T., R. W. Mitchell, and M. L. Boccia. 1994. *Self-Awareness in Animals and Humans: Developmental Perspectives.* Cambridge: Cambridge University Press.

Patterson, F. G. P., and R. H. Cohn. 1994. Self-recognition and self-awareness in lowland gorillas. In S. T. Parker, R. W. Mitchell, and M. L. Boccia, eds., *Self-Awareness in Animals and Humans: Developmental Perspectives,* pp. 273–290. Cambridge: Cambridge University Press.

Pavelka, M. S. McDonald- 1993. *Monkeys of the Mesquite: The Social Life of the South Texas Snow Monkey.* Dubuque, Iowa: Kendall/Hunt.

Perry, S., and L. Rose. 1994. Begging and transfer of coati meat by white-faced capuchin monkeys, *Cebus capucinus. Primates* 35:409–415.

Piaget, J. 1962 (1928). *The Moral Judgment of the Child.* New York: Crowell-Collier and Macmillan.

Pilleri, G. 1984. Epimeletic behaviour in Cetacea: Intelligent or instinctive? In G. Pilleri, ed., *Investigations on Cetacea,* pp. 30–48. Vammala, Finland: Vammalan Kirjapaino Oy.

Pinker, S. 1994. Is there a gene for compassion? *New York Times Book Review,* Sept. 25.

Porter, J. W. 1977. Pseudorca stranding. *Oceans* 4:8–14.

Povinelli, D. J. 1987. Monkeys, apes, mirrors and minds: The evolution of self-awareness in primates. *Human Evolution* 2:493–507.

—— 1994. How to create self-recognizing gorillas (but don't try it on macaques). In S. T. Parker, R. W. Mitchell, and M. L. Boccia, eds., *Self-Awareness in Animals and Humans: Developmental Perspectives,* pp. 291–300. Cambridge: Cambridge University Press.

Povinelli, D. J., S. T. Boysen, and K. E. Nelson. 1990. Inferences about guessing and knowing by chimpanzees *(Pan troglodytes). Journal of Comparative Psychology* 104:203–210.

Povinelli, D. J., K. E. Nelson, and S. T. Boysen. 1992. Comprehension of social role reversal by chimpanzees:Evidence of empathy? *Animal Behaviour* 43:633–640.

Povinelli, D. J., A. B. Rulf, and M. A. Novak. 1992. Role reversal by rhesus monkeys, but no evidence of empathy. *Animal Behaviour* 44:269–281.

Power, M. 1991. *The Egalitarians: Human and Chimpanzee.* Cambridge: Cambridge University Press.

Premack, D., and G. Woodruff. 1978. Does the chimpanzee have a theory of mind? *Behavioral and Brain Sciences* 1:515–526.

Pryor, K. 1975. *Lads before the Wind.* New York: Harper and Row.

Rasa, O. A. E. 1976. Invalid care in the dwarf mongoose *(Helogale undulata rufula). Zeitschrift für Tierpsychologie* 42:337–342.

——— 1979. The effects of crowding on the social relationships and behaviour of the dwarf mongoose *(Helogale undulata rufula). Zeitschrift für Tierpsychologie* 49:317–329.

Rawls, J. 1972. *A Theory of Justice.* Oxford: Oxford University Press.

Regal, P. J. 1990. *The Anatomy of Judgment.* Minneapolis: University of Minnesota Press.

Reinhardt, V., R. Dodsworth, and J. Scanlan. 1986. Altruistic interference shown by the alpha-female of a captive group of rhesus monkeys. *Folia primatologica* 46:44–50.

Ren, R., et al. 1991. The reconciliation behavior of golden monkeys *(Rhinopithecus roxellanae roxellanae)* in small breeding groups. *Primates* 32:321–327.

van Rhijn, J., and R. Vodegel. 1980. Being honest about one's intentions: An evolutionarily stable strategy for animal conflicts. *Journal of Theoretical Biology* 85:623–641.

Richards, R. J. 1987. *Darwin and the Emergence of Evolutionary Theories of Mind and Behavior.* Chicago: University of Chicago Press.

Rousseau, J. 1965 (1755). *Discours sur l'origine et les fondements de l'inégalité parmi les hommes.* Paris: Gallimard.

Ruse, M. 1986. *Taking Darwin Seriously: A Naturalistic Approach to Philosophy.* Oxford: Blackwell.

——— 1988. Response to Williams: Selfishness is not enough. *Zygon* 23:413–416.

Sabater Pi, J., et al. 1993. Behavior of bonobos *(Pan paniscus)* following their capture of monkeys in Zaire. *International Journal of Primatology* 14:797–804.

Sackin, S., and E. Thelen. 1984. An ethological study of peaceful associative outcomes to conflict in preschool children. *Child Development* 55:1098–1102.

Saffire, W. 1990. The bonding market. *New York Times Magazine,* June 24.

Sahlins, M. D. 1965. On the sociology of primitive exchange. In M. Banton, ed., *The Relevance of Models for Social Anthropology,* pp. 139–236. London: Tavistock.

Salk, L. 1973. The role of the heartbeat in the relations between mother and infant. *Scientific American* 228:24–29.

Salzinger, K. 1990. B. F. Skinner 1904–1990. *American Psychological Association Observer* 3:1–4.

Savage-Rumbaugh, E. S., et al. 1986. Spontaneous symbol acquisition and communicative use by two pygmy chimpanzees. *Journal of Experimental Psychology* 115:211–235.

Scanlon, C. E. 1986. Social development in a congenitally blind infant rhesus macaque. In R. G. Rawlins and M. J. Kessler, eds., *The Cayo Santiago Macaques,* pp. 94–109. Albany: State University of New York Press.

van Schaik, C. P. 1983. Why are diurnal primates living in groups? *Behaviour* 87:120–144.

———— 1989. The ecology of social relationships amongst female primates. In V. Standen and R. A. Foley, eds., *Comparative Socioecology: The Behavioral Ecology of Humans and Other Mammals,* pp. 195–218. Oxford: Blackwell.

van Schaik, C. P., and M. A. van Noordwijk. 1985. Evolutionary effect of the absence of felids on the social organization of the macaques on the island of Simeulue (*Macaca fascicularis fusca,* Miller 1903). *Folia primatologica* 44:138–147.

Schama, S. 1987. *The Embarrassment of Riches: An Interpretation of Dutch Culture in the Golden Age.* New York: Knopf.

Scott, J. F. 1971. *Internalization of Norms: A Sociological Theory of Moral Commitment.* Englewood Cliffs, N.J.: Prentice-Hall.

Scott, J. P. 1958. *Animal Behavior.* Chicago: University of Chicago Press.

Seton, E. T. 1907. *The Natural History of the Ten Commandments.* New York: Scribners.

Seville Statement on Violence. 1986. Middletown, Conn.: Wesleyan University.

Seyfarth, R. M. 1981. Do monkeys rank each other? *Behavioral and Brain Sciences* 4:447–448.

———— 1983. Grooming and social competition in primates. In R. Hinde, ed., *Primate Social Relationships: An Integrated Approach,* pp. 182–190. Sunderland, Mass.: Sinauer.

Seyfarth, R. M., and D. L. Cheney. 1988. Empirical tests of reciprocity theory: Problems in assessment. *Ethology and Sociobiology* 9:181–188.

Sherif, M. 1966. *In Common Predicament: Social Psychology of Intergroup Conflict and Cooperation.* Boston: Houghton Mifflin.

Sigg, H., and J. Falett. 1985. Experiments on respect of possession in hamadryas baboons *(Papio hamadryas). Animal Behaviour* 33:978–984.

Silk, J. B. 1992a. The origins of caregiving behavior. *American Journal of Physical Anthropology* 87:227–229.

———— 1992b. The patterning of intervention among male bonnet macaques: Reciprocity, revenge, and loyalty. *Current Anthropology* 33:318–325.

Simon, H. A. 1990. A mechanism for social selection and successful altruism. *Science* 250:1665–68.

Singer, P. 1976. *Animal Liberation.* London: Jonathan Cape.

———— 1981. *The Expanding Circle: Ethics and Sociobiology.* New York: Farrar, Straus and Giroux.

Skinner, B. F. 1990. Can psychology be a science of mind? *American Psychologist* 45:1206–10.

Small, M. F. 1990. Social climber: Independent rise in rank by a female Barbary macaque *(Macaca sylvanus). Folia primatologica* 55:85–91.

Smetana, J. G., M. Killen, and E. Turiel. 1991. Children's reasoning about interpersonal and moral conflicts. *Child Development* 62:629–644.

Smith, A. 1937 (1759). *A Theory of Moral Sentiments.* New York: Modern Library.

—— 1982 (1776). *An Inquiry into the Nature and Causes of the Wealth of Nations.* Indianapolis: Liberty Classics.

Smuts, B. B. 1985. *Sex and Friendship in Baboons.* New York: Aldine.

—— 1987. Gender, aggression, and influence. In B. B. Smuts et al., eds., *Primate Societies,* pp. 400–412. Chicago: University of Chicago Press.

Smuts, B. B., and J. M. Watanabe. 1990. Social relationships and ritualized greetings in adult male baboons *(Papio cynocephalus anubis). International Journal of Primatology* 11:147–172.

Sober, E. 1984. *The Nature of Selection: Evolutionary Theory in Philosophical Focus.* Cambridge, Mass.: MIT Press.

—— 1988. What is evolutionary altruism? *Canadian Journal of Philosophy* 14:75–99.

Stanford, C. B., et al. 1994a. Patterns of predation by chimpanzees on red colobus monkeys in Gombe National Park, 1982–1991. *American Journal of Physical Anthropology* 94:213–228.

—— 1994b. Hunting decisions in wild chimpanzees. *Behaviour* 131:1–18.

von Stephanitz, M. 1950 (1918). *The German Shepherd Dog.* 8th rev. ed. Augsburg: Verein für Deutsche Schäferhunde.

Strier, K. B. 1992a. Causes and consequences of nonaggression in the woolly spider monkey, or muriqui *(Brachyteles arachnoides).* In J. Silverberg and J. P. Gray, eds., *Aggression and Peacefulness in Humans and Other Primates,* pp. 100–116. New York: Oxford University Press.

—— 1992b. *Faces in the Forest: The Endangered Muriqui Monkeys of Brazil.* New York: Oxford University Press.

Strum, S. C. 1987. *Almost Human: A Journey into the World of Baboons.* New York: Random House.

Sugiyama, Y. 1984. Population dynamics in wild chimpanzees at Bossou, Guinea, between 1976 and 1983. *Primates* 25:391–400.

—— 1988. Grooming interactions among adult chimpanzees at Bossou, Guinea, with special reference to social structure. *International Journal of Primatology* 9:393–408.

Tannen, D. 1990. *You Just Don't Understand: Women and Men in Conversation.* New York: Ballantine.

Taylor, C. E., and M. T. McGuire. 1988. Reciprocal altruism: Fifteen years later. *Ethology and Sociobiology* 9:67–72.

Teleki, G. 1973a. Group response to the accidental death of a chimpanzee in Gombe National Park, Tanzania. *Folia primatologica* 20:81–94.

────── 1973b. *The Predatory Behavior of Wild Chimpanzees*. Lewisburg, Penn.: Bucknell University Press.

Temerlin, M. K. 1975. *Lucy: Growing up Human*. Palo Alto, Calif.: Science and Behavior Books.

Thierry, B., and J. R. Anderson. 1986. Adoption in anthropoid primates. *International Journal of Primatology* 7:191–216.

Thompson, N. S. 1976. My descent from the monkey. In P. Bateson and P. Klopfer, eds., *Perspectives in Ethology*, pp. 221–230. New York: Plenum.

Tinklepaugh, O. L. 1928. An experimental study of representative factors in monkeys. *Journal of Comparative Psychology* 8:197–236.

Tobin, J. J., D. Y. H. Wu, and D. H. Davidson. 1989. *Preschool in Three Cultures*. New Haven: Yale University Press.

Todes, D. 1989. *Darwin without Malthus: The Struggle for Existence in Russian Evolutionary Thought*. New York: Oxford University Press.

Tokida, E., et al. 1994. Tool-using in Japanese macaques: Use of stones to obtain fruit from a pipe. *Animal Behaviour* 47:1023–30.

Tomasello, M., A. Kruger, and H. Ratner. 1993. Cultural learning. *Behavioral and Brain Sciences* 16:495–552.

Trivers, R. L. 1971. The evolution of reciprocal altruism. *Quarterly Review of Biology* 46:35–57.

────── 1974. Parent-offspring conflict. *American Zoologist* 14:249–264.

Trumler, E. 1974. *Hunde Ernst Genommen: Zum Wesen und Verständnis ihres Verhaltens*. Munich: Piper.

Turnbull, C. M. 1972. *The Mountain People*. New York: Touchstone.

Varley, M., and D. Symmes. 1966. The hierarchy of dominance in a group of macaques. *Behaviour* 27:54–75.

Vehrencamp, S. A. 1983. A model for the evolution of despotic versus egalitarian societies. *Animal Behaviour* 31:667–682.

Visalberghi, E., and D. M. Fragaszy. 1990. Do monkeys ape? In S. Parker and K. Gibson, eds., *"Language" and Intelligence in Monkeys and Apes: Comparative Developmental Perspectives*, pp. 247–273. Cambridge: Cambridge University Press.

Vogel, C. 1985. Evolution und Moral. In H. Maier-Leibnitz, ed., *Zeugen des Wissens*, pp. 467–507. Mainz: Hase und Koehler.

────── 1988. Gibt es eine natürliche Moral? *Civis* 1:65–78.

Vollmer, P. J. 1977. Do mischievous dogs reveal their "guilt"? *Veterinary Medicine Small Animal Clinician* 72:1002–5.

de Waal, F. B. M. 1977. The organization of agonistic relations within two captive groups of Java-monkeys *(Macaca fascicularis)*. *Zeitschrift für Tierpsychologie* 44:225–282.

—— 1982. *Chimpanzee Politics: Power and Sex among Apes*. London: Jonathan Cape.

—— 1984. Sex-differences in the formation of coalitions among chimpanzees. *Ethology and Sociobiology* 5:239–255.

—— 1986a. The brutal elimination of a rival among captive male chimpanzees. *Ethology and Sociobiology* 7:237–251.

—— 1986b. Deception in the natural communication of chimpanzees. In R. W. Mitchell and N. S. Thompson, eds., *Deception: Human and Nonhuman Deceit*, pp.221–244. Albany: State University of New York Press.

—— 1986c. Integration of dominance and social bonding in primates. *Quarterly Review of Biology* 61:459–479.

—— 1987. Tension regulation and nonreproductive functions of sex among captive bonobos *(Pan paniscus)*. *National Geographic Research* 3:318–335.

—— 1989a. *Peacemaking among Primates*. Cambridge, Mass.: Harvard University Press.

—— 1989b. Dominance "style" and primate social organization. In V. Standen and R. A. Foley, eds., *Comparative Socioecology: The Behavioural Ecology of Humans and Other Mammals,* pp. 243–264. Oxford: Blackwell.

—— 1989c. The myth of a simple relation between space and aggression in captive primates. *Zoo Biology Supplement* 1:141–148.

—— 1989d. Food sharing and reciprocal obligations among chimpanzees. *Journal of Human Evolution* 18:433–459.

—— 1990. Do rhesus mothers suggest friends to their offspring? *Primates* 31:597–600.

—— 1991a. Complementary methods and convergent evidence in the study of primate social cognition. *Behaviour* 118:297–320.

—— 1991b. The chimpanzee's sense of social regularity and its relation to the human sense of justice. *American Behavioral Scientist* 34:335–349.

—— 1992a. Aggression as a well-integrated part of primate social relationships: Critical comments on the Seville Statement on Violence. In J. Silverberg and J. P. Gray, eds., *Aggression and Peacefulness in Humans and Other Primates*, pp. 37–56. New York: Oxford University Press.

—— 1992b. Appeasement, celebration, and food sharing in the two *Pan* species. In T. Nishida et al., eds., *Topics in Primatology,* vol. 1, *Human Origins*, pp. 37–50. Tokyo: University of Tokyo Press.

—— 1992c. A social life for chimpanzees in captivity. In J. Erwin and J. C. Landon, eds., *Chimpanzee Conservation and Public Health: Environments for the Future*, pp. 83–87. Rockville, Md.: Diagnon/Bioqual.

—— 1992d. Intentional deception in primates. *Evolutionary Anthropology* 1:86–92.

—— 1993a. Reconciliation among primates: A review of empirical evidence and unresolved issues. In W. A. Mason and S. P. Mendoza, eds., *Primate Social Conflict*, pp. 111–144. Albany: State University of New York Press.

———— 1993b. Sex differences in chimpanzee (and human) behavior: A matter of social values? In M. Hechter, L. Nadel, and R. Michod, eds., *The Origin of Values*, pp. 285–303. New York: Aldine de Gruyter.

———— 1993c. Co-development of dominance relations and affiliative bonds in rhesus monkeys. In M. E. Pereira and L. A. Fairbanks, eds., *Juvenile Primates: Life History, Development, and Behavior*, pp. 259–270. New York: Oxford University Press.

———— 1994. The chimpanzee's adaptive potential: A comparison of social life under captive and wild conditions. In R. W. Wrangham et al., eds., *Chimpanzee Cultures*, pp. 243–260. Cambridge, Mass.: Harvard University Press.

de Waal, F. B. M., and F. Aureli. Forthcoming. Reconciliation, consolation, and a possible cognitive difference between macaque and chimpanzee. In A. E. Russon, K. A. Bard, and S. T. Parker, eds., *Reaching into Thought: The Minds of the Great Apes*. Cambridge: Cambridge University Press.

de Waal, F. B. M., and D. L. Johanowicz. 1993. Modification of reconciliation behavior through social experience: An experiment with two macaque species. *Child Development* 64:897–908.

de Waal, F. B. M., and L. M. Luttrell. 1985. The formal hierarchy of rhesus monkeys: An investigation of the bared-teeth display. *American Journal of Primatology* 9:73–85.

———— 1986. The similarity principle underlying social bonding among female rhesus monkeys. *Folia primatologica* 46:215–234.

———— 1988. Mechanisms of social reciprocity in three primate species: Symmetrical relationship characteristics or cognition? *Ethology and Sociobiology* 9:101–118.

———— 1989. Toward a comparative socioecology of the genus *Macaca*: Intergroup comparisons of rhesus and stumptail monkeys. *American Journal of Primatology* 10:83–109.

de Waal, F. B. M., and R. Ren. 1988. Comparison of the reconciliation behavior of stumptail and rhesus macaques. *Ethology* 78:129–142.

de Waal, F. B. M., and A. van Roosmalen. 1979. Reconciliation and consolation among chimpanzees. *Behavioral Ecology and Sociobiology* 5:55–66.

de Waal, F. B. M., and D. Yoshihara. 1983. Reconciliation and re-directed affection in rhesus monkeys. *Behaviour* 85:224–241.

de Waal, F. B. M., L. M. Luttrell, and M. E. Canfield. 1993. Preliminary data on voluntary food sharing in brown capuchin monkeys. *American Journal of Primatology* 29:73–78.

de Waal, F. B. M., et al. In press. Behavioral retardation in a macaque with autosomal trisomy and aging mother. *American Journal on Mental Retardation.*

Walker, L. J. 1984. Sex differences in the development of moral reasoning: A critical review. *Child Development* 55:677–691.

Walker Leonard, J. 1979. A strategy approach to the study of primate dominance behaviour. *Behavioural Processes* 4:155–172.

Walters, J. 1980. Interventions and the development of dominance relationships in female baboons. *Folia primatologica* 34:61–89.

Weiss, R. F., et al. 1971. Altruism is rewarding. *Science* 171:1262–63.

Wendt, H. 1971. *From Ape to Adam.* London: Thames and Hudson.

Westergaard, G. C., C. W. Hyatt, and W. D. Hopkins. 1994. The responses of bonobos *(Pan paniscus)* to mirror-image stimulation: Evidence of self-recognition. *Journal of Human Evolution* 9:273–279.

Whiten, A. 1991. *Natural Theories of Mind: Evolution, Development and Simulation of Everyday Mindreading.* Oxford: Blackwell.

Whiten, A., and R. W. Byrne. 1988. Tactical deception in primates. *Behavioral and Brain Sciences* 11:233–273.

Whiten, A., and R. Ham. 1992. On the nature and evolution of imitation in the animal kingdom: Reappraisal of a century of research. In P. J. B. Slater et al. eds., *Advances in the Study of Behavior,* vol. 21, pp. 239–283. New York: Academic Press.

Wickler, W. 1981 (1971). *Die Biologie der Zehn Gebote: Warum die Natur für uns kein Vorbild ist.* Munich: Piper.

Wilkinson, G. S. 1984. Reciprocal food sharing in the vampire bat. *Nature* 308:181–184.

Williams, G. C. 1966. *Adaptation and Natural Selection.* Princeton: Princeton University Press.

——— 1988. Reply to comments on "Huxley's evolution and ethics in sociobiological perspective." *Zygon* 23:437–438.

——— 1989. A sociobiological expansion of "Evolution and Ethics." In *Evolution and Ethics,* pp. 179–214. Princeton: Princeton University Press.

Wilson, D. S. 1983. The group selection controversy: History and current status. *Annual Review of Ecology and Systematics* 14:159–187.

Wilson, D. S., and E. Sober. 1994. Reintroducing group selection to the human behavioral sciences. *Behavioral and Brain Sciences* 17:585–654.

Wilson, E. O. 1975. *Sociobiology: The New Synthesis.* Cambridge, Mass.: Belknap Press, Harvard University Press.

Wilson, J. Q. 1993. *The Moral Sense.* New York: Free Press.

Wispé, L. 1991. *The Psychology of Sympathy.* New York: Plenum.

Woodburn, J. 1982. Egalitarian societies. *Man* 17:431–451.

Wrangham, R. W. 1979. On the evolution of ape social systems. *Social Science Information* 18:335–368.

———— 1980. An ecological model of female-bonded primate groups. *Behaviour* 75:262–300.

Wrangham, R. W., A. P. Clark, and G. Isabirye-Basuta. 1992. Female social relationships and social organization of Kibale Forest chimpanzees. In T. Nishida et al., eds., *Topics in Primatology,* vol. 1, *Human Origins,* pp. 81–98. Tokyo: University of Tokyo Press.

Wrangham, R. W., et al. 1994. *Chimpanzee Cultures.* Cambridge, Mass.: Harvard University Press.

Wright, R. 1994. *The Moral Animal: The New Science of Evolutionary Psychology.* New York: Pantheon.

Yerkes, R. M. 1925. *Almost Human.* New York: Century.

———— 1941. Conjugal contrasts among chimpanzees. *Journal of Abnormal Social Psychology* 36:175–199.

Yerkes, R. M., and A. W. Yerkes. 1929. *The Great Apes: A Study of Anthropoid Life.* New Haven: Yale University Press.

———— 1935. Social behavior in infrahuman primates. In *A Handbook of Social Psychology,* pp. 973–1033. Worcester, Mass.: Clark University Press.

Zahn-Waxler, C., B. Hollenbeck, and M. Radke-Yarrow. 1984. The origins of empathy and altruism. In M. W. Fox and L. D. Mickley, eds., *Advances in Animal Welfare Science,* pp. 21–39. Washington, D.C.: Humane Society of the United States.

Zahn-Waxler, C., et al. 1992. Development of concern for others. *Developmental Psychology* 28:126–136.

ACKNOWLEDGMENTS

This book reflects more than two decades of primate research—and a lifelong love of animals. As a child, I caught and kept sticklebacks, raised jackdaws, and bred mice, always fascinated with the details of their behavior. It was only later that I discovered that some of these creatures also happened to be the favorites of two renowned ethologists, Konrad Lorenz and Niko Tinbergen. No wonder I identified with these men and their groundbreaking work.

I was educated in the Dutch ethological tradition by eminent scientists such as Gerard Baerends, Jaap Kruijt, Piet Wiepkema, and particularly Jan van Hooff, under whose supervision I conducted my first primate studies. My six years with the chimpanzees at the Arnhem Zoo are still a highlight. Core ideas of the present book were developed during this period, if not in full-fledged form then at least in utero. I am grateful to everyone involved, from the zoo director, Anton van Hooff, to the many students of the University of Utrecht.

Most material for this book was collected in the 1980s at the Wisconsin Regional Primate Research Center in Madison. I did make a few expeditions to San Diego and Atlanta to study bonobos and chimpanzees, but Wisconsin was the home base where I had rhesus, stump-tailed, and capuchin monkeys. Two people were absolutely crucial: Robert Goy, the director who invited me to the center, and

Lesleigh Luttrell, who assisted me with great dedication in my many projects. Peter Judge and RenMei Ren joined my team as scientists, and students of the University of Wisconsin contributed to the research: Kim Bauers, Eloise Canfield, Charles Chaffin, Karen Friedlen, Lisa Jeannotte, Denise Johanowicz, Brenda Miller, Katherine Offutt, Amy Parish, and Kurt Sladky. I very much enjoyed my interactions with fellow ethologists and primatologists in Madison, especially Charles Snowdon and Karen Strier.

In 1991 I moved to Atlanta, Georgia, to join the scientific staff of the Yerkes Regional Primate Research Center as well as the faculty of the Department of Psychology at Emory University. The principal reason was my desire to continue working with chimpanzees, an understudied species in relation to the interest it deserves. The Yerkes Primate Center and its director, Frederick King, have been most generous in helping me set up a research operation that today includes a team of approximately fifteen enthusiastic people. Of these, let me mention in particular technicians Lisa Parr and Michael Seres and postdoctoral associate Filippo Aureli; they have made invaluable contributions to the research reported in this book.

Atlanta and nearby Athens, Georgia, probably have the world's largest concentration of both nonhuman primates and primatologists; the intellectual environment has proved most stimulating. Colleagues in the psychology department have taught me to think critically about a number of issues with which, as a zoologist, I had little familiarity.

Over the years my research has been generously supported by two regional primate research centers, as well as by funding from the National Science Foundation, the National Institutes of Mental Health, the National Institutes of Health, the Harry Frank Guggenheim Foundation, and Emory University.

The genesis of this book was a 1989 meeting in Monterey, California, of the Gruter Institute for Law and Behavioral Research. For many years Margaret Gruter, a legal scholar with a keen interest in evolutionary theory, has brought together international experts to explore the interface between law and biology. This particular meeting, entitled "Biology, Law, and the Sense of Justice," first made me realize how the phenomena that I was studying in nonhuman primates could shed light on the origin of justice and morality in our own species. By assembling scholars of widely disparate disciplines, Margaret has rendered all of us a tremendous service. The influence of her institute will reverberate throughout academic life for a long while.

Because morality is such a broad topic, I invited comments on drafts of the manuscript from many scholars—from economists to philosophers, and from cognitive psychologists to fellow ethologists and biologists. I learned much from their reflections and sometimes was given references to significant literature that I had not been aware of. Much of this was done with the convenience of the Internet. Naturally, not every reviewer agreed with me, and not every disagreement could be resolved. I remain fully responsible for the opinions expressed in this book.

Commentators on portions of the manuscript and contributors of observations or citations include Otto Adang, Richard Alexander, Filippo Aureli, Kim Bard, Christopher Boehm, Sue Boinsky, Gordon Burghardt, Josep Call, Richard Connor, Marina Cords, Hank Davis, Lee Dugatkin, Robert Frank, David Goldfoot, Harold Gouzoules, Christine Johnson, Peter Judge, Melanie Killen, Bruce Knauft, Stella Launy, Elizabeth Lloyd, Rudolf Makkreel, Anne McGuire, Ulric Neisser, Toshisada Nishida, Ronald Noë, Mary McDonald Pavelka, Michael Pereira, Susan Perry, Daniel Povinelli, John Robinson, Philippe Rochat, Sue Savage-Rumbaugh, Laele Sayigh, Carel van Schaik, Jeanne Scheurer, Rachel Smolker, Elliott Sober, Volker Sommer, Emanuela Cenami Spada, Karen Strier, Hiroyuki Takasaki, Ichirou Tanaka, Michael Tomasello, Robert Trivers, Peter Verbeek, Elisabetta Visalberghi, Kim Wallen, Andrew Whiten, and David Sloan Wilson.

In addition, I was fortunate to receive comments from four exceptionally incisive readers who went over the manuscript intensively, each from his or her own perspective. These were Barbara Smuts, well versed in the theoretical controversies that run through this book as well as an expert on baboon behavior; Lesleigh Luttrell, who as my assistant of twelve years knows every monkey subject by name and background, and in addition has excellent editorial judgment; Michael Fisher, my editor and champion at Harvard University Press; and my wife, Catherine Marin, who read every draft as it came off my printer. The detailed feedback from these four persons helped enormously in shaping the book's content and form. I also thank Vivian Wheeler for the thoroughness with which she edited the final text.

The majority of the photographs are mine, taken on black-and-white Kodak film with a Minolta camera and lenses ranging from 50 to 500 millimeters. Robert Dodsworth and Frank Kiernan, photographers at the Wisconsin and Yerkes primate centers, respectively, did much of the darkroom work. For additional illustrations I thank

Donna Bierschwale, Christophe Boesch, Gordon Burghardt, Susan Meier, Emil Menzel, Cynthia Moss, Ronald Noë, and RenMei Ren.

Catherine created the environment that made it possible for me to complete this challenging yet enjoyable task, providing support, warmth, and humor at home despite a demanding academic career. I could have no greater pleasure, therefore, than to dedicate this book to her.

INDEX

For ease of identification, main entries that are names of individual primates are italicized.

68, 230n32; attributional capacity and, 75, 237n12; dominance patterns and, 126, 235n8; group life and, 168; deception and, 225n6. *See also* Bonobos; Chimpanzees; Gorillas; Orangutans

Aping. *See* Imitation

Aquatic ape theory, 227n19

Ardrey, Robert, 195

Arnhem Zoo chimpanzee colony (Netherlands): conflict mediation at, 32–33, 129, 130; lethal aggression in, 38; deception in, 44, 76–77; responses to death at, 55; imitation at, 72; perspective-taking at, 83; social regulation in, 89; teasing techniques at, 114; sharing in, 143; group life in, 164, 168; hierarchies and, 172–173; responses to crowding at, 197–198; orphans at, 227n17; female in control role at, 241n36

Atlanta, 13, 19–20, 203

Attachment: among whales, 42–43; succorant behavior and, 52–53, 80, 81; individualism and, 167; as ingredient of morality, 211. *See also* Friendship; Kinship; Learned adjustment; Succorant behavior

Attributional capacity, 73–78, 237n12

Atwood, Margaret, 239n26

Aureli, Filippo, 60, 159, 182, 200, 228n24

Azalea, 49–50, 69, 81, 113

Baboons: friendship among, 19; responses to death, 54–55; infant privileges among, 112; teasing among, 114; sharing and, 142; alliances and, 155–156; mother-infant relations among, 177; greetings between males, 191–193; respect for possession among, 245n18

Bacon, Ellis, 225n8

Badcock, C. R., 238n18

Baker, K., 248n6

Bats, 20–21, 144

Batson, Daniel, 87–88, 233n52

Bauers, Kim, 91

Bear cubs, 225n8

Beatle, 26

Behavioral sink, 194–195, 198

Behaviorism, 34–35, 37, 62

Belle, 73–74

Bercovitch, Frederick, 155

Berkson, G., 226n12

Bernstein, Irwin, 113, 128

Biami (Papuan tribe in New Guinea), 66–67, 71

Biology and morality relation, 1, 10–13, 36–38

Biomedical research, 215–216, 254n4

Birds, 12, 35, 36, 43. *See also* Ravens

Bischof-Köhler, D., 230n34

Bjorn, 233n48

Blindness, 51–52, 226n12

Blurton Jones, Nicholas, 152

Blushing, 114–117

Boehm, Christopher, 128, 129–130, 132, 240n33

Boesch, Christophe, 58, 140–141, 142

Boesch, Hedwige, 140–141, 142

Boinsky, Sue, 84, 171

Bonaventura, Saint, 97

Bonding, as term, 19

Bonnet monkeys, 246n24

Bonobos: perspective taking and, 82; sex-for-food exchange among, 154, 245n20; reconciliations among, 176, 177, 229n31; self-recognition and, 229n32; sharing and, 243n11

Bowlby, John, 174

Boysen, Sarah, 74

Boz, 192

Brain structures. *See* Neurobiology

Breggin, Peter, 194

Breuggeman, J. A., 231n37

Bridging behavior, 122–123

Broughton, J. M., 238n21

Buraya (tribe), 128

Burghardt, Gordon, 37, 64

Byrne, Richard, 78

Caine, N., 225n6

Calhoun, John, 194–195, 198, 252n33

Call, Josep, 200

Calvinism, 17

Camper, Petrus, 65

Canetti, Elias, 20

Canids. *See* Dogs; Wolves

Cannibalism, 242n7

Captive *vs.* wild, 168–169. *See also* Wild chimpanzees

Capuchin monkeys, 81–82, 84, 143–144, 148, 150, 189–190

Carpenter, Edmund, 66–67
Cats, domestic, 92, 154–155
Cat's Eye (Atwood), 239n26
Causation, proximate *vs.* ultimate, 222n27, 238n18
Cavalieri, Paola, 214–215
Ceausescu, Nicolea, 98
Celebration: and food-sharing, 140, 151–152, 244n17; reconciliation and, 205
Cerebral palsy, 50–51
Cetaceans. *See* Dolphins; Whales
Chagnon, Napoleon, 161
Charlesworth, William, 188
Cheating, 159
Cheney, Dorothy, 75, 204, 232n41, 245n22
Chicago Academy of Sciences Conference (1991), 141
Children, human: perspective taking and, 74; distinctions among rule violations and, 105; peacemaking skills and, 180–181, 249n18; conflicts among, 247n4
Chimpanzee Politics (de Waal), 77, 97, 129, 143
Chimpanzees: birth process among, 19–20; empathy among, 19–20, 56–57, 58, 85, 210, 233n48; alliances among, 28, 127, 157–162, 172–173, 205, 241n33; intergroup conflict among, 30; mediation among, 32–33, 129–130; aggression among, 38, 124, 228nn23,24; lethal aggression among, 38, 194; deception and, 44, 75–78, 149–150, 237n12; responses to death, 54, 55–57; responses to injuries, 58, 61–62, 228n23; consolation and, 60–61, 70; self-awareness and, 67–68, 70–71; imitation and, 72–73; tool use by, 72–73, 210; attributional capacities of, 73–78, 237n12; perspective taking and, 73–78, 83; lack of compassion among, 83–84; social rules and, 89, 91–92, 93, 95; reciprocity among, 97, 150–154, 157–162, 245n19; dominance relations among, 99, 100, 102, 103–104, 235n10, 239n28, 241n33; reconciliation and, 103–104, 164–165, 176, 200, 204–205, 239n28, 252n35; teasing among, 114, 203;

sex differences among, 124, 239n28; dominance style among, 126–127; hunting by, 140–141, 142; food-sharing and, 140–142, 147, 149–150, 245n19; revenge and, 157–162; indignation and, 160; group life and, 168–169, 247n5; reunions among, 174; political murder among, 223n34; facial discrimination by, 248n13; natural habitat of, 253n36. *See also* Wild chimpanzees
Circle, moral: sharing as central to, 144–146; as expanding, 146, 212–214
Coco, 228n19
Coe, Christopher, 110
Cofeeding, 244n16
Cognitive abilities in apes, 77–78, 82–83, 210. *See also* Deception; Empathy; Imitation; Mirror self-recognition; Perspective taking
Cognitive empathy, 48, 69–70, 80, 81, 211
Cognitive ethology, 3
Cognitive psychology, 228n25
Colmenares, Fernando, 112, 191, 251n24
Colobus monkeys, 140
Comforting contact, 46, 60–61, 178
Common enemy, 22–23, 29–31, 127, 170–171
Community concern, 31–34, 203–208, 211; defined, 207
Compassion, lack of: among nonhuman primates, 83–85; among tribal humans, 85–86
Competition: natural selection and, 11–12; *vs.* cooperation, as evolutionary principle, 21–23, 27; hierarchies and, 103; food-sharing and, 140; moral behavior and, 161–162; sexual, among the muriquis, 201; constraints on, 206–207, 254n41. *See also* Conflict
Confiscation of weapons, 32–33, 205–206
Conflict: intergroup, 29–31, 252n39; within the group, 29–34, 199–200, 223n34, 240n32, 247n4; types of outcomes of, 99–100; sex differences in relation to, 123–124, 239n26, 241n36; weaning, 177, 187–189,

251n21; in human marriage, 184–
186; as negotiation, 187; indirect,
238n26. *See also* Aggression; Competition; Mediation; Reconciliation
Conflict arbitration. *See* Control role
Conflict resolution, 30–31; community
concern and, 31–34; nature of aggression and, 166, 170; as ingredient of
morality, 211. *See also* Mediation;
Reconciliation
Conscience: autonomous, 91–92; building of, 109; evolution of, 117,
237n17. *See also* Guilt; Shame
Consciousness. *See* Self-awareness
Consolation, 60–61, 70, 230n35
Contact: comforting, 46, 60–61, 178;
invalid care and, 80; male baboon
greetings and, 191–193; between ri-
vals, 231, 235n10. *See also* Grooming
Control role, 128–132, 241n36
Cooperation: *vs.* competition, as evolu-
tionary principle, 21–23, 27; distin-
guished from reciprocal altruism, 24;
formal hierarchies and, 103, 104. *See
also* Alliance formation
Copying. *See* Imitation
Cords, Marina, 181
Crawford, Meredith, 147, 148
Creationism, 228n25
Crime, and primate research, 193–203
Cripples. *See* Handicapped group mem-
bers
Critical anthropomorphism, 64
Cross-species empathy, 214–216,
249n16, 254n4
Cross-species morality, 84–85
Cross-species sympathy, 42, 224n3
Crowding: behavioral sink hypothesis
and, 194–195, 198; aggression as re-
sponse to, 194–197, 199; primate
coping skills and, 197–201
Cruelty to animals. *See* Cross-species
empathy
Cultural variation, 36, 201, 248n6
Culture, primate, 178–180, 210, 254n2
Cuvier, Baron, 222n30

Damasio, Hanna, 217
D'Amato, M. R., 148
Dandy, 76, 77
Darley, John, 87–88
Dart, R. A., 247n3

Darwin, Charles, 1, 11–12, 22, 23, 40,
116, 117, 220n9
Dasser, Verena, 204
Davis, Hank, 105–106, 107, 111
Dawe, Helen, 249n18
Dawkins, Richard, 6, 14–15, 16,
220n12
Deafness, in chimpanzees, 122
Death, responses to, 53–56
Deception: chimpanzees and, 44, 75–
78, 149–150, 237n12; sympathy en-
trapment and, 44; blushing and, 116;
spontaneous, 149–150; capacity for,
224n6; attributional capacity and,
237n12
Democracy, 241n33
Descartes, René, 63, 214
Dettwyler, K. A., 7–8
Development. *See* Moral development
Dewey, John, 209
Dingell, John, 194
Diogenes, 65
Disney, Walt, 81
Dittus, Wolfgang, 58, 196
"Divide-and-rule" strategy, 172–173
Dogs: succorant behavior and, 41; cog-
nitive empathy and, 48; social rules
and, 93, 94–95; internalization of
rules and, 106–109; efficacy of pun-
ishment with, 107–108, 237n11; dif-
ferences among breeds and, 108
Dolphins: altruistic behavior among,
12; sympathy and, 40, 42, 43, 59; re-
action to danger, 171–172; birth
among, 221n18; self-awareness and,
230n36
Domestic abuse, 240n32
Dominance drive, 98–100
Dominance style: despotic, 125, 130;
variability of, 125–126
Double-holding, 101–102
"Drummer Girl" (horse), 190–191
Dualisms, 228n25
Dugatkin, Lee, 71
Dutch culture, 126, 127, 194, 240n31

East Germany, 240n32
Edwards, C. P., 238n24
Egalitarianism: *vs.* despotism, 125,
130; dominance relations and, 127–
128, 132, 144, 240n33; sharing and,
137; muriquis and, 201

Ik (tribe), 85–86, 213, 242n3
IM (two-headed snake), 37
Imanishi, Kinji, 254n2
Imitation, 19–20, 71–73, 180, 231nn37,39. *See also* Learning
Impartial spectator, 33–34, 109
Imprinting, 35, 36
Indignation, 159–160. *See also* Moralistic aggression
Indirect reciprocity, 33, 237n17
Indirect revenge, 159
Individual and society, tensions between, 171–172, 206–208. *See also* Self-sacrifice
Individual model of aggression, 164, 173
Infants. *See* Mother-infant bond
Injuries, responses to, 56–60
Innate releasers, 80–81
Intentions: human moral judgment and, 15, 73; gap between consequences and, 28–29; of animals, 46–47, 73, 96–97, 113, 232n41; communication of, 191–193, 203
Interests, conflux of, 27
Intergroup conflict, 29–31
Internalization of rules, 105–111
Invisible hand metaphor, 28
Isaac, Glynn, 133, 146

Jacoby, Susan, 160
Jakie, 83
Janson, Charles, 145
Japanese culture: drinking etiquette in, 136; conflict resolution in, 180–181
Japanese macaques: treatment of the handicapped, 6–7, 8, 9, 50–51, 227n14; matrilines and, 8, 9, 187; friendship among, 26; aggression in, 158–159, 196–197
Jigokudani Park, 6, 8, 26, 48
Jimoh, 91–92, 131–132
Johan (jackdaw), 134
Johanowicz, Denise, 179
Johnston, Kay, 119–120
Joubert, Beverly, 53
Joubert, Dereck, 53
Judge, Peter, 199, 200, 204
Justice: human social regulation and, 125; control role and, 131; revenge and, 160–162, 184; individual autonomy and, 167

Kalind, 82
Kalunde, 143
Kant, Immanuel, 10, 87
Kanzi, 210
Kaplan, Hillard, 137
Kaufman, Charles, 54
Kennedy, Edward, 194
Kevin, 174–175
Keyes, Randall, 52
Killen, M., 247n4
Killer Ape myth, 38, 167, 247n3
Kindchenschema, 81
Kin selection, 12, 15, 135, 144, 155, 212. *See also* Group selection; Reciprocal altruism
Kinship: in macaque society, 8; nepotism and, 18; conflict mediation and, 36; reciprocal altruism and, 135; triadic reconciliation and, 203. *See also* Matrilines
Kissing, origin of, 43
Knauft, B. M., 241n33
Knowledge of relationships, 100–101, 203, 235n7
Koestler, Arthur, 102
Kohlberg, Lawrence, 92–93, 105
Köhler, Wolfgang, 68, 85
Koko, 210, 230n32
Kortlandt, A., 229n31
Krazy Kat (cartoon), 184, 185
Krebs, Dennis, 25
Krom, 72, 83
Kropotkin, Petr, 21–23, 27, 29, 162
Kummer, Hans, 172, 181, 245n18, 251n24
Kurland, Jeffrey, 9, 249n20

Language: human acquisition of, 36; role in morality, 45, 79; anthropomorphism and, 63–64; research on ape use of, 111, 210
Langurs. *See* Golden monkeys
Larson, Gary, 217
Law-and-order mentality, 93
Leadership, and dominance, 128–132
Learned adjustment, 48–53, 56–57, 211
Learning: age specificity and, 36, 223n31; black box approach to, 62; of rules, 94–95; of conflict-resolution skills, 178–180, 249n17; active teaching and, 211; observational, 231n39;

cultural transmission and, 254n2. *See also* Imitation
Lee, Richard, 138
Left-side cradling bias, 78–79
Leveling mechanisms, 128, 131, 241n33
Lever, Janet, 118–119, 123
Lewin, R., 223n32
Lieberman, Philip, 45
Lifeboat ethics, 213, 217
Long-tailed macaques, 163–164, 170–171, 181, 182, 200
Lopez, Barry, 94
Lorenz, Konrad, 35, 38, 106, 163, 164, 221n21, 223n33, 228n23, 247n3
Loretta, 82
Lucy, 57, 148
Luit, 97, 129, 143, 157, 172–173, 227n17, 235n10
Luttrell, Lesleigh, 46
Lux, Kenneth, 219n7
Lying. *See* Deception

Macaques: treatment of the handicapped, 6–7, 8, 9, 50–51, 227n14; matrilines and, 8, 9, 26, 187; birth process among, 20; friendship among, 26–27; cleaning of wounds among, 58; consolation and, 60–61, 70, 230n35; self-awareness and, 70; social hierarchy and, 101–102; dominance hierarchies among, 109–111, 152; sharing and, 152; alliances and, 156–157, 158–159; reciprocity and, 156–157, 158–159; reconciliation and, 176; aggression in, 196–197; deception among, 225n6. *See also* Apes, differences between monkeys and; *specific types of monkeys*
Machiavellian strategies, 234n6
Madame Bee, 147–148
Madrid Zoo, 112, 191
Mai, 13–14, 19–20, 160
Malthus, Thomas, 11, 13, 195, 219n7
Mama, 174, 241n36
de Mandeville, Bernard, 27, 28, 251n24
Mango, 107, 108
Marie, 226n9
Marriage, 184–186
Maslow, Abraham, 99, 126–127, 234n6
Masters, Roger, 169
Maternal behavior, 89–91
Mating: sneaky, 76–77, 91–92, 109–

111; altruistic behavior and, 133–135, 137, 242n5; muriquis and, 201; as motivation for hunting, 245n20
Matrilines: in macaque society, 9, 26, 155; confrontations between, 155; triadic reconciliation and, 203; infighting and, 250n20. *See also* Kinship
Matsuzawa, T., 248n13
Maynard Smith, J., 254n41
McDonald-Pavelka, M. S.. *See* Pavelka, M. S. McDonald-
McGregor, 242n47
McGrew, W. C., 140, 244n16
McGuire, Michael, 198
Meat, status of, 138–140, 146. *See also* Predation
Mediation, 31–33, 205–206; in human societies, 128; among primates, 128–132, 197; sex differences and, 241n36. *See also* Control role; Reciprocity, indirect; Triadic reconciliation
Meier, Susan, 107
Menzel, Emil, 73–74, 78
Mercer, Philip, 79
Midgley, Mary, 10, 14, 163, 220n12
Miller, Brenda, 158–159
Milton, Katharine, 137
Mirror self-recognition: first reactions of humans and, 66–67; in apes, 67–68, 229n32; as test of self-awareness, 68–71
Mixed-currency exchanges, 155–156
Mongooses, 80, 144
Monkeys. *See* Apes, differences between monkeys and; Baboons; Macaques; *specific types of monkeys*
Moral ability, 36
Moral contract. *See* Social contract
Moral development: stages of, 10, 36, 92–93, 105; development of succorant tendencies and, 44–47; primate rule acquisition and, 111–114; sex differences in, 119–122; conflict-resolution skills and, 177; weaning conflict and, 188. *See also* Children, human
Moralistic aggression, 159, 211, 234n4
Morality: philosophers and, 10, 146, 218; biological view of, 10–13, 36–38; conditions for evolution of, 29–31, 34, 39; cultural variability of, 36; sympathy-based *vs.* rule-based, 119;

Price, G. R., 254n41
Primate order, evolutionary tree for, 4
Promiscuous altruism, 214
Prosimians. *See* Ring-tailed lemurs
Proudhon, Pierre-Joseph, 214
Provisioning, 253n36
Puist, 76, 97
Punishment: assertive reconciliation
and, 104; after the fact, 107–108,
237n11; rule acquisition in juveniles
and, 112; revenge and, 157–159, 160

Quid pro quo. *See* Reciprocity

Rank. *See* Formal dominance
Rasa, Anne, 80, 198
Rats, 105–106, 107, 195
Ravens, 133–134
Rawls, John, 161, 167
Reciprocal altruism, 12, 24–27; characteristics of, 24; distinguished from cooperation, 24; link between morality
and, 136–144; expectations and, 153–
154; revenge and, 157–159, 160; evolution of conscience and, 237n17.
See also Kin selection; Trivers, Robert
Reciprocity: similarity principle and,
26; among hunter-gatherers, 137–
138; experimental evidence for, 150–
154; alliances and, 154–159; as ingredient of morality, 211; indirect,
237n17; centralized, 243n10; forms
of, in human societies, 243n10. *See
also* Reciprocal altruism; Revenge
Reconciliation: as label, 18–19; mediation and, 32–33; assertive, 103–104;
sex differences in patterns of, 123–
124, 239n28; nature of aggression
and, 164–166; defined, 176; proto-
cols for, 176; learning of, 177–182;
strategic, 181; maladaptive, 186;
crowded conditions and, 200,
252n35; teasing and, 203; triadic,
204; research methods and, 249n14
Reductionism, 17, 102, 220n12
Reite, M., 225n6
Relational model of aggression, 173–
176, 186, 193
Relationship, value of, and conflict resolution, 181, 200. *See also* Relational
model
Religion, 17, 30, 117

Ren, RenMei, 31–32
Repression theory, 238n18
Reproductive success. *See* Mating
Reputation, 115–116, 237n17
Retribution. *See* Revenge
Reunion euphoria, 174–176
Reunions, 174–176. *See also* Greetings
Revenge: direct, 157–158, 246n24; indirect, 159; justice and, 160–162, 184;
as ingredient of morality, 211
Rhesus monkeys: friendship among, 26–
27, 101; empathy among, 45–46, 47,
61, 69; treatment of the handicapped
and, 49–50, 51–52; food-sharing
and, 93, 95; social hierarchies and,
99, 101–102, 104, 109–111; double-
holding among, 101–102; reconciliation and, 104, 177–180, 200,
249n17; punishment among, 112–
113; rule learning among, 112–113;
dominance style among, 126–127;
levels of aggression among, 198–200;
imitation and, 231n37; nurturant behavior among, 240n29; conflict mediation and, 241n36; intergroup relations among, 253n39. *See also*
Macaques
Rhett, 203
van Rhijn, J., 254n41
Right and wrong, 2, 39, 105, 123, 201
Rights: sex differences in human morality and, 119–122; individualism and,
166–167; of animals, 214–216. *See
also* Cross-species empathy
Ring-tailed lemurs, 59–60
Rita, 47
Rix, 56
Robinson, John, 81–82
Rock, 73–74
Rockefeller, John D., 11
Role-taking, 69–70, 73–78
van Roosmalen, Angeline, 200
Ropey, 26, 61, 101–102
Rose, Lisa, 145
Rosenblum, Leonard, 54, 110
Roth, E., 196–197
Rousseau, Jean-Jacques, 166–167, 196
Rule-based morality, 119
Rules: group sanctions and, 89, 91–92;
descriptive *vs.* prescriptive, 90–92,
211; acquisition of, 94–95, 111–114;
intentionality of, 96–97; internaliza-